Project Planning & Design (PPD)

ARE 5.0 Mock Exam
(Architect Registration Examination)

ARE 5.0 Overview, Exam Prep Tips,
Hot Spots, Case Studies, Drag-and-Place,
Solutions and Explanations

Gang Chen

ArchiteG®, Inc.
Irvine, California

Project Planning & Design (PPD) ARE 5.0 Mock Exam (Architect Registration Examination): ARE 5.0 Overview, Exam Prep Tips, Hot Spots, Case Studies, Drag-and-Place, Solutions and Explanations
Copyright © 2017   Gang Chen
V1.0
Cover Photo © 2017   Gang Chen

Copy Editor: Penny L Kortje

All Rights Reserved.
No part of this book may be transmitted or reproduced by any means or in any form, including electronic, graphic, or mechanical, without the express written consent of the publisher or author, except in the case of brief quotations in a review.

ArchiteG, Inc.
http://www.ArchiteG.com

ISBN: 978-1-61265-029-6

PRINTED IN THE UNITED STATES OF AMERICA

What others are saying about *ARE Mock Exam series* …

"Great study guide…"
"This was a great resource supplement to my other study resources. I appreciated the mock exam questions the most, and the solutions offer an explanation as to why the answer is correct. I will definitely check out his other ARE exam resources!

UPDATE: Got my PASS Letter!"
—**Sean Primeaux**

"Tried everything 4 times before reading this book and PASSED!"
"I had failed this exam 4 times prior to getting this book…I had zero clue as to what I was doing wrong. I read Ballast, Kaplan and random things on the forum but for the life of me couldn't pin point where I was missing it until I read THIS BOOK! Gang did an excellent job…I remember going through the ramp and reading Gang's book and saying Ohhhh like 4 or 5 times. I read his book several times until I became comfortable with the information. I went in on test day and it was a breeze. I remember walking out of there thinking I couldn't believe I struggled so much before. The tips in here are priceless! I strongly recommend this book…"
—**hendea1**

"Add this to your ARE study"
"This was a very helpful practice exam and discussion. I really appreciated the step-by-step review of the author's approach... As I studied it last before taking the test, Gang Chen's book probably made the difference for me."
—**Dan Clowes ("XLine")**

"Good supplemental mock exam"
"I found the mock exam to be very helpful, all of the answers are explained thoroughly and really help you understand why it is correct...Also the introduction and test taking tips are very helpful for new candidates just starting the ARE process."
—**Bgrueb01**

"Essential Study Tool"
"I have read the book and found it to be a great study guide for myself. Mr. Gang Chen does such a great job of helping you get into the right frame of mind for the content of the exam. Mr. Chen breaks down the points on what should be studied and how to improve your chances of a pass with his knowledge and tips for the exam and practice vignettes.

I highly recommend this book to anyone…it is an invaluable tool in the preparation for the exam as Mr. Chen provides a vast amount of knowledge in a very clear, concise, and logical matter."
—**Luke Giaccio**

"Wish I had this book earlier"
"…The questions are written like the NCARB questions, with various types...check all that apply, fill in the blank, best answer, etc. The answer key helpfully describes why the correct answer is correct, and why the incorrect answers are not. Take it from my experience, at half the cost of

other mock exams, this is a must buy if you want to pass..."
—**Domiane Forte ("Vitruvian Duck")**

"This book did exactly like the others said."
"This book did exactly like the others said. It is immensely helpful with the explanation... There are so many codes to incorporate, but Chen simplifies it into a methodical process. Bought it and just found out I passed. I would recommend."
—**Dustin**

"It was the reason I passed."
"This book was a huge help. I passed the AREs recently and I felt this book gave me really good explanations for each answer. It was the reason I passed."
—**Amazon Customer**

"Great Practice Exam"
"… For me, it was difficult to not be overwhelmed by the amount of content covered by the Exam. This Mock Exam is the perfect tool to keep you focused on the content that matters and to evaluate what you know and what you need to study. It definitely helped me pass the exam!!"
—**Michael Harvey ("Harv")**

"One of the best practice exams"
"Excellent study guide with study tips, general test info, and recommended study resources. Hands down one of the best practice exams that I have come across for this exam. Most importantly, the practice exam includes in depth explanations of answers. Definitely recommended."
—**Taylor Cupp**

"Great Supplement!!"
"This publication was very helpful in my preparation for my BS exam. It contained a mock exam, followed by the answers and brief explanations to the answers. I would recommend this as an additional study material for this exam."
—**Cynthia Zorrilla-Canteros ("czcante")**

"Fantastic! "
"When I first began to prepare for this exam; the number of content areas seemed overwhelming and daunting at best. However, this guide clearly dissected each content area into small management components. Of all the study guides currently available for this test - this exam not only included numerous resources (web links, you tube clips, etc..), but also the sample test was extremely helpful. The sample test incorporated a nice balance of diagrams, calculations and general concepts - this book allowed me to highlight any "weak" content areas I had prior to the real exam. In short - this is an awesome book!"
—**Rachel Casey (RC)**

Dedication

To my parents, Zhuixian and Yugen,
my wife, Xiaojie, and my daughters,
Alice, Angela, Amy, and Athena.

Disclaimer

Project Planning & Design (PPD) ARE 5.0 Mock Exam (Architect Registration Examination) provides general information about Architect Registration Examination. The book is sold with the understanding that neither the publisher nor the authors are providing legal, accounting, or other professional services. If legal, accounting, or other professional services are required, seek the assistance of a competent professional firm.

The purpose of this publication is not to reprint the content of all other available texts on the subject. You are urged to read other materials, and tailor them to fit your needs.

Great effort has been taken to make this resource as complete and accurate as possible. However, nobody is perfect and there may be typographical errors or other mistakes present. You should use this book as a general guide and not as the ultimate source on this subject. If you find any potential errors, please send an e-mail to:
info@ArchiteG.com

Project Planning & Design (PPD) ARE 5.0 Mock Exam (Architect Registration Examination) is intended to provide general, entertaining, informative, educational, and enlightening content. Neither the publisher nor the author shall be liable to anyone or any entity for any loss or damages, or alleged loss or damages, caused directly or indirectly by the content of this book.

ArchiteG®, Green Associate Exam Guide®, GA Study®, and GreenExamEducation® are registered trademarks owned by Gang Chen.

ARE®, Architect Registration Examination® are registered trademarks owned by NCARB.

If you do not wish to be bound by the above, you may return this book to the publisher for a full refund.

Legal Notice

ARE Mock Exam series by ArchiteG, Inc.

Time and effort is the most valuable asset of a candidate. How to cherish and effectively use your limited time and effort is the key of passing any exam. That is why we publish the ARE Mock Exam series to help you to study and pass the ARE exams in the shortest time possible. We have done the hard work so that you can save time and money. We do not want to make you work harder than you have to. To save your time, we use a *standard* format for all our ARE 5.0 Mock Exam books, so that you can quickly skip the *identical* information you have already read in other books of the series, and go straight to the *unique* "meat and potatoes" portion of the book.

The trick and the most difficult part of writing a good book is to turn something that is very complicated into something that is very simple. This involves researching and really understanding some very complicated materials, absorbing the information, and then writing about the topic in a way that makes it very easy to understand. To succeed at this, you need to know the materials very well. Our goal is to write books that are clear, concise, and helpful to anyone with a seventh-grade education.

Do not force yourself to memorize a lot of numbers. Read through the numbers a few times, and you should have a very good impression of them.

You need to make the judgment call: If you miss a few numbers, you can still pass the exam, but if you spend too much time drilling these numbers, you may miss out on the big pictures and fail the exam.

The existing ARE practice questions or exams by others are either way too easy or way over-killed. They do NOT match the real ARE 5.0 exams at all.

We have done very comprehensive research on the official NCARB guides, many related websites, reference materials, and other available ARE exam prep materials. We match our mock exams as close as possible to the NCARB samples and the real ARE exams instead. Some readers had failed an ARE exam two or three times before, and they eventually passed the exam with our help.

All our books include a complete set of questions and case studies. We try to mimic the real ARE exams by including the same number of questions, using a similar format, and asking the same type of questions. We also include detailed answers and explanations to our questions.

There is some extra information on ARE overviews and exam-taking tips in Chapter One. This is based on NCARB *and* other valuable sources. This is a bonus feature we included in each book because we want our readers to be able to buy our ARE mock exam books together or individually. We want you to find all necessary ARE exam information and resources at one place and through our books.

All our books are available at
http://www.GreenExamEducation.com

How to Use This Book

We suggest you read *Project Planning & Design (PPD) ARE 5.0 Mock Exam (Architect Registration Examination)* at least three times:

Read once and cover Chapter One and Two, the Appendixes, the related *free* PDF files, and other resources. Highlight the information you are not familiar with.

Read twice focusing on the highlighted information to memorize. You can repeat this process as many times as you want until you master the content of the book.

After reviewing these materials, you can take the mock exam, and then check your answers against the answers and explanations in the back, including explanations for the questions you answer correctly. You may have answered some questions correctly for the wrong reason. Highlight the information you are not familiar with.

Like the real exam, the mock exam will continue to use **multiple choice, check-all-that-apply,** and **quantitative fill-in-the-blank**. There are also three new question types: **hot spots, case studies,** and **drag-and-place**.

Review your highlighted information, and take the mock exam again. Try to answer 100% of the questions correctly this time. Repeat the process until you can answer all the questions correctly.

PPD is one of the most difficult ARE divisions because many PPD questions require calculations. This book includes most if not all the information you need to do the calculations, as well as step-by-step explanations. After reading this book, you will greatly improve your ability to deal with the real ARE PPD calculations, and have a great chance of passing the exam on the first try.

Take the mock exam at least two weeks before the real exam. You should definitely NOT wait until the night before the real exam to take the mock exam. If you do not do well, you will go into panic mode and NOT have enough time to review your weaknesses.

Read for the final time the night before the real exam. Review ONLY the information you highlighted, especially the questions you did not answer correctly when you took the mock exam for the first time.

This book is very light so you can easily carry it around. These features will allow you to review the graphic vignette section whenever you have a few minutes.

The Table of Contents is very detailed so you can locate information quickly. If you are on a tight schedule you can forgo reading the book linearly and jump around to the sections you need.

All our books, including "ARE Mock Exams Series" and "LEED Exam Guides Series," are available at

GreenExamEducation.com

Check out FREE tips and info at **GeeForum.com**, you can post your questions for other users' review and responses.

Table of Contents

Dedication..5

Disclaimer...7

ARE Mock Exam Series by ArchiteG, Inc...9

How to Use This Book..11

Table of Contents..13

Chapter One   Overview of Architect Registration Examination (ARE)

    A. First Thing First: Go to the Website of your Architect Registration Board and Read all the Requirements of Obtaining an Architect License in your Jurisdiction..17

    B. Download and Review the Latest ARE Documents at the NCARB Website..17
       1. Important links to the FREE and official NCARB documents
       2. A detailed list and brief description of the FREE PDF files that you can download from NCARB
- ARE 5.0 Credit Model
- ARE 5.0 Guidelines
- NCARB Education Guidelines
- Architectural Experience Program (AXP) Guidelines
- Certification Guidelines
- ARE 5.0 Related FAQs (Frequently Asked Questions)
- Your Guide to ARE 5.0
- ARE 5.0 Handbook
- ARE 5.0 Test Specification
- ARE 5.0 Prep Videos
- The Burning Question: Why Do We need ARE anyway?
- Defining Your Moral Compass
- Rules of Conduct

    C. The Intern Development Program (IDP)/Architectural Experience Program (AXP)..22
       1. What is IDP? What is AXP?
       2. Who qualifies as an intern?

D. Overview of Architect Registration Examination(ARE)..22
1. How to qualify for the ARE?
2. How to qualify for an architect license?
3. What is the purpose of ARE?
4. What is NCARB's rolling clock?
5. How to register for an ARE exam?
6. How early do I need to arrive at the test center?
7. Exam Format & Time
    - Practice Management (PcM)
    - Project Management (PjM)
    - Programming & Analysis (PA)
    - Project Planning & Design (PPD)
    - Project Development & Documentation (PDD)
    - Construction & Evaluation (CE)
8. How are ARE scores reported?
9. Is there a fixed percentage of candidates who pass the ARE exams?
10. When can I retake a failed ARE division?
11. How much time do I need to prepare for each ARE division?
12. Which ARE division should I take first?
13. ARE exam prep and test-taking tips
14. Strategies for passing ARE exams on the first try
    - Find out how much you already know and what you should study
    - Cherish and effectively use your limited time and effort
    - Do NOT stretch your exam prep process too long
    - Resist the temptation to read many books and limit your time and effort to read only a few selected books or a few sections of books in details
    - Think like an architect.
15. ARE exam preparation requires short-term memory
16. Allocation of your time and scheduling
17. Timing of review: the 3016 rule; memorization methods, tips, suggestions, and mnemonics
18. The importance of good and effective study methods
19. The importance of repetition: read this book <u>at least</u> three times
20. The importance of a routine
21. The importance of short, frequent breaks and physical exercise
22. A strong vision and a clear goal
23. English system (English or inch-pound units) vs. metric system (SI units)
24. Codes and standards used in this book
25. Where can I find study materials on architectural history?

Chapter Two   **Project Planning & Design (PPD) Division**

    A. General Information..37
        1. Exam content
        2. Official exam guide and reference index for the Project Planning & Design (PPD) division

    B. The Most Important Documents/Publications for PPD Division of the ARE Exam.38
        1. *Official NCARB list of formulas and references for the Project Planning & Design (PPD) division with our comments and suggestions*
          Publications; Codes; AIA Contract Documents
        2. *Manual of Steel Construction: Allowable Stress Design*
        3. *Steel Construction Manual*
        4. FREE information on truss and beam diagrams
        5. The FREE PDF file of FEMA publication number 454 (FEMA454), *Designing for Earthquakes: A Manual for Architects*
        6. The FREE PDF file of *Wind Design Made Simple*
        7. *Building at Risk*
        8. Construction Specifications Institute (CSI) MasterFormat & *Building Construction*

Chapter Three   **ARE Mock Exam for Project Planning & Design (PPD) Division**

    A. Multiple-Choice (MC) 45

    B. Case Study..88

Chapter Four   **ARE Mock Exam Solutions for Project Planning & Design (PPD) Division**

    A. Mock Exam Answers and Explanations: Multiple-Choice (MC)................97

    B. Mock Exam Answers and Explanations: Case Study.............................134

Appendixes
    A. List of Figures..139
    B. Official reference materials suggested by NCARB................................142
        1. Resources Available While Testing
        2. Typical Beam Nomenclature
        3. Formulas Available While Testing
        4. Common Abbreviations
        5. General NCARB reference materials for ARE
        6. Official NCARB reference materials matrix
        7. Extra Study Materials
    C. Other reference materials..159

    D. Some Important Information about Architects and the Profession of Architecture..160
    E. AIA Compensation Survey..166
    F. So … You would Like to Study Architecture..167

Back Page Promotion
    A. ARE Mock Exam series (GreenExamEducation.com)
    B. LEED Exam Guides series (GreenExamEducation.com)
    C. *Building Construction* (ArchiteG.com)
    D. *Planting Design Illustrated*

Index

Chapter One

Overview of the Architect Registration Examination (ARE)

A. **First Thing First: Go to the Website of your Architect Registration Board and Read all the Requirements of Obtaining an Architect License in your Jurisdiction**
See the following link:
http://www.ncarb.org/Getting-an-Initial-License/Registration-Board-Requirements.aspx

B. **Download and Review the Latest ARE Documents at the NCARB Website**

1. **Important links to the FREE and official NCARB documents**
NCARB launched ARE 5.0 on November 1, 2016. ARE 4.0 will continue to be available until June 30, 2018.

ARE candidates who started testing in ARE 4.0 can choose to "self-transition" to ARE 5.0. This will allow them to continue testing in the version that is most suitable for them. However, **once a candidate transitions to ARE 5.0, s/he cannot transition back to ARE 4.0**.

The current version of the Architect Registration Examination (ARE 5.0) includes six divisions:

- Practice Management (PcM)
- Project Management (PjM)
- Programming & Analysis (PA)
- Project Planning & Design (PPD)
- Project Development & Documentation (PDD)
- Construction & Evaluation (CE)

All ARE divisions continue to use **multiple choice, check-all-that-apply,** and **quantitative fill-in-the-blank**. The new exams include three new question types: **hot spots, case studies,** and **drag-and-place**.

There is a tremendous amount of valuable information covering every step of becoming an architect available free of charge at the NCARB website:
http://www.ncarb.org/

For example, you can find guidance about architectural degree programs accredited by the National Architectural Accrediting Board (NAAB), NCARB's Architectural Experience Program (AXP) formerly known as Intern Development Program (IDP), and licensing requirements by state. These documents explain how you can qualify to take the Architect Registration Examination.

We find the official ARE 5.0 Guidelines, ARE 5.0 Handbook, and ARE 5.0 Credit Model extremely valuable. See the following link:
http://www.ncarb.org/ARE/ARE5.aspx

You should start by studying these documents.

2. **A detailed list and brief description of the FREE PDF files that you can download from NCARB**
   The following is a detailed list of the FREE PDF files that you can download from NCARB. They are listed in order based on their importance.

   - All **ARE 5.0** information can be found at the following links**:**
     http://www.ncarb.org/ARE/ARE5.aspx
     http://blog.ncarb.org/2016/November/ARE5-Study-Materials.aspx
   - The **ARE 5.0 Credit Model** is one of the most important documents, and tells you the easiest way to pass the ARE by taking selected divisions from ARE 4.0 and ARE 5.0.

ARE5.0:	Practice Management	Project Management	Programming & Analysis	Project Planning & Design	Project Development & Documentation	Construction & Evaluation
ARE 4.0:						
Construction Documents & Services	●	●			●	●
Programming Planning & Practice	●	●	●			
Site Planning & Design			●	●		
Building Design & Construction Systems				●	●	
Structural Systems				●	●	
Building Systems				●	●	
Schematic Design				●		

As shown in matrix above, if you are taking both ARE 4.0 and ARE 5.0, you can pass the ARE exams by taking only five divisions in total. To complete the ARE, your goal is to select and pass exams from both versions which cover all sixteen dots in matrix above. The quickest potential options are as follows:

a. You can take the following five divisions to pass the ARE:
   ARE 4.0
   - Construction Documents & Services
   - Programming Planning & Practice
   - Site Planning & Design

   ARE 5.0
   - Project Planning & Design
   - Project Development & Documentation

OR

b. You can take the following five divisions to pass the ARE:
   ARE 4.0
   - Construction Documents & Services
   - Programming Planning & Practice

   ARE 5.0
   - Programming & Analysis
   - Project Planning & Design
   - Project Development & Documentation

- **ARE 5.0 Guidelines** includes extremely valuable information on the ARE overview, NCARB, registration (licensure), architectural education requirements, the Architectural Experience Program (AXP), establishing your eligibility to test, scheduling an exam appointment, taking the ARE, receiving your score, retaking the ARE, the exam format, scheduling, and links to other FREE NCARB PDF files. You need to read this <u>at least twice</u>.

- **NCARB Education Guidelines** contains information on education requirements for initial licensure and for NCARB certification, satisfying the education requirement, foreign-educated applicants, the education alternative to NCARB certification, the Education Evaluation Services for Architects (EESA), the Education Standard, and other resources.

- **Architectural Experience Program (AXP) Guidelines** includes information on AXP overview, getting started and creating your NCARB record, experience areas and tasks, documenting your experience through hours, documenting your experience through a portfolio, and the next steps. You need to read this document <u>at least twice</u>. The information is valuable.

NCARB renamed the **Intern Development Program (IDP)** as **Architectural Experience Program (AXP)** in June 2016. Most of NCARB's 54-member boards have adopted the AXP as a prerequisite for initial architect licensure. Therefore, you should be familiar with the program.

The AXP application fee is $100. This fee includes one free transmittal of your Record for initial registration and keeps your Record active for the first year. After the initial year, there is an annual renewal fee required to maintain an active Record until you become registered. The cost is currently $85 each year. The fees are subject to change, and you need to check the NCARB website for the latest information.

There are two ways to meet the AXP requirements. The **first method** is **reporting hours**. Most candidates will use this method. You will need to document at least 3,740 required hours under the six different experience areas to complete the program. A minimum of 50% of your experience must be completed under the supervision of a qualified architect.

The following chart lists the hours required under the six experience areas:

Experience Area	Hours Required
Practice Management	160
Project Management	360
Programming & Analysis	260
Project Planning & Design	1,080
Project Development & Documentation	1,520
Construction & Evaluation	360
Total	**3,740**

Your experience reports will fall under one of **two experience settings**:
• **Setting A**: Work performed for an architecture firm.
• **Setting O**: Experiences performed outside an architecture firm.

You must earn at least **1,860 hours** in experience **setting A**.

Your AXP experience should be reported to NCARB at least every six months and logged within two months of completing each reporting period (the **Six-Month Rule**).

The **second method** to meet AXP requirements is to create an **AXP Portfolio**. This new method is for experienced design professionals who put their licensure on hold and allows you to prove your experience through the preparation of an online portfolio.

To complete the AXP through the **second method**, you will need to meet ALL the AXP criteria through the portfolio. In other words, you cannot complete the experience requirement through a combination of the **AXP portfolio** and **reporting hours**.

See the following link for more information on AXP:
http://www.ncarb.org/Experience-Through-Internships.aspx

- **Certification Guidelines** by NCARB (Skimming through this should be adequate. You should also forward a copy of this PDF file to your IDP supervisor.)

See the following link which contains resources for supervisors and mentors:

http://www.ncarb.org/Experience-Through-Internships/Supervisors-and-Mentors/Resources-for-Supervisor-and-Mentors.aspx

- **ARE 5.0 Related FAQs (Frequently Asked Questions)**: Skimming through this should be adequate.

- **Your Guide to ARE 5.0** includes information on understanding the basics of ARE 5.0, new question types, taking the test, making the transition, getting ARE 5.0 done, and planning your budget. The document also contains FAQs, and links for more information. You need to read this document at least twice. The information is valuable.

- **ARE 5.0 Handbook** contains an ARE overview, detailed information for each ARE division, and ARE 5.0 references. This handbook explains what NCARB expects you to know so that you can pass the ARE exams. ARE 5.0 uses either **Understand/Apply (U/A)** or **Analyze/Evaluate (A/E)** to designate the appropriate cognitive complexity of each objective, but *avoids* the use of **"Remember,"** the lowest level of cognitive complexity (CC), or **"Create,"** the highest level of CC.

    This handbook has some sample questions for each division. The real exam is like the samples in this handbook.

    Tips:
    - *ARE 5.0 Handbook has about 170 pages. To save time, you can just read the generic information at the front and back portion of the handbook, and focus on the ARE division(s) you are currently studying for. As you progress in your testing you can read the applicable division that you are studying for. This way the content will always be fresh in your mind.*
    - *You need to read this document at least three times. The information is valuable.*

- **ARE 5.0 Test Specification** identifies the ARE 5.0 division structure and defines the major content areas, called **Sections**; the measurement **Objectives**; and the percentage of content coverage, called **Weightings**. This document specifies the scope and objectives of each ARE division, and the percentage of questions in each content category. You need to read this document at least twice. The information is valuable, and is the base of all ARE exam questions.

- **ARE 5.0 Prep Videos** include one short video for each division. These videos give you a very good basic introduction to each division, including sample questions and answers, and explanations. You need to watch each video at least three times. See the following link:
    http://blog.ncarb.org/2016/November/ARE5-Study-Materials.aspx

- **The Burning Question: Why Do We Need ARE Anyway?** (We do not want to give out a link for this document because it is too long and keeps changing. You can Google it with its title. Skimming through this document should be adequate.)

- **Defining Your Moral Compass** (You can Google it with its title plus the word "NCARB." Skimming through this document should be adequate.)

- **Rules of Conduct** is available as a FREE PDF file at:
  http://www.ncarb.org/
  (Skimming through this should be adequate.)

C. The Intern Development Program (IDP)/Architectural Experience Program (AXP)

1. What is IDP? What is AXP?
IDP is a comprehensive training program jointly developed by the National Council of Architectural Registration Boards (NCARB) and the American Institute of Architects (AIA) to ensure that interns obtain the necessary skills and knowledge to practice architecture <u>independently</u>. NCARB renamed the **Intern Development Program (IDP)** as **Architectural Experience Program (AXP)** in June 2016.

2. Who qualifies as an intern?
Per NCARB, if an individual meets one of the following criteria, s/he qualifies as an intern:
a. Graduates from NAAB-accredited programs
b. Architecture students who acquire acceptable training prior to graduation
c. Other qualified individuals identified by a registration board

D. Overview of the Architect Registration Examination (ARE)

1. How to qualify for the ARE?
A candidate needs to qualify for the ARE via one of NCARB's member registration boards, or one of the Canadian provincial architectural associations.

Check with your Board of Architecture for specific requirements.

For example, in California, a candidate must provide verification of a minimum of <u>five</u> years of education and/or architectural work experience to qualify for the ARE.

Candidates can satisfy the five-year requirement in a variety of ways:

- Provide verification of a professional degree in architecture through a program that is accredited by NAAB or CACB.

  OR
- Provide verification of at least five years of educational equivalents.

  OR
- Provide proof of work experience under the direct supervision of a licensed architect

See the following link:
http://www.ncarb.org/ARE/Getting-Started-With-the-ARE/Ready-to-Take-the-ARE-Early.aspx

2. **How to qualify for an architect license?**
   Again, each jurisdiction has its own requirements. An individual typically needs a combination of about <u>eight</u> years of education and experience, as well as passing scores on the ARE exams. See the following link:
   http://www.ncarb.org/Reg-Board-Requirements

   For example, the requirements to become a licensed architect in California are:
   - Eight years of post-secondary education and/or work experience as evaluated by the Board (including at least one year of work experience under the direct supervision of an architect licensed in a U.S. jurisdiction or two years of work experience under the direct supervision of an architect registered in a Canadian province)
   - Completion of the Architectural Experience Program (AXP)
   - Successful completion of the Architect Registration Examination (ARE)
   - Successful completion of the California Supplemental Examination (CSE)

   California does NOT require an accredited degree in architecture for examination and licensure. However, many other states do.

3. **What is the purpose of ARE?**
   The purpose of ARE is NOT to test a candidate's competency on every aspect of architectural practice. Its purpose is to test a candidate's competency on providing professional services to protect the <u>health, safety, and welfare</u> of the public. It tests candidates on the <u>fundamental</u> knowledge of pre-design, site design, building design, building systems, and construction documents and services.

   The ARE tests a candidate's competency as a "specialist" on architectural subjects. It also tests her abilities as a "generalist" to coordinate other consultants' works.

   You can download the exam content and references for each of the ARE divisions at the following links:
   http://www.ncarb.org/ARE/ARE5.aspx
   http://www.ncarb.org/are/40/StudyAids.html

4. **What is NCARB's rolling clock?**
   a. Starting on January 1, 2006, a candidate MUST pass ALL ARE sections within five years. A passing score for an ARE division is only valid for five years, and a candidate has to retake this division if she has NOT passed all divisions within the five-year period.

   b. Starting on January 1, 2011, a candidate who is authorized to take ARE exams MUST take at least one division of the ARE exams within five years of the authorization.

Otherwise, the candidate MUST apply for the authorization to take ARE exams from an NCARB member board again.

These rules were created by the **NCARB's rolling clock** resolution and passed by NCARB council during the 2004 NCARB Annual Meeting.

ARE 4.0 division expiration dates per the Rolling Clock will remain the same for the transition to ARE 5.0.

5. **How to register for an ARE exam?**
   See the instructions in the new ARE guideline at the following link:
   http://www.ncarb.org/ARE/ARE5.aspx

6. **How early do I need to arrive at the test center?**
   Be at the test center at least 30 minutes BEFORE your scheduled test time, OR you may lose your exam fee.

7. **Exam format & time**
   All ARE divisions are administered and graded by computer. Their time allowances are as follows:

Division	Number of Questions	Test Duration	Appointment Time
Practice Management	80	2 hr 45 min	3 hr 30 min
Project Management	95	3 hr 15 min	4 hr
Programming & Analysis	95	3 hr 15 min	4 hr
Project Planning & Design	120	4 hr 15 min	5 hr
Project Development & Documentation	120	4 hr 15 min	5 hr
Construction & Evaluation	95	3 hr 15 min	4 hr
Total Time:		21 hr	25 hr 30 min

NCARB suggests you to arrive at the test center a minimum of 30 minutes before your scheduled appointment. You can have one flexible 15-minute break for each division. That is why the appointment time is 45 minutes longer than the actual test time for each division.

Practice Management has 80 questions and NCARB allows you 2 hours and 45 minutes to complete the exam, so you should spend an average of (2x60+45)/80=165/80= 2.06 minutes on each question.

Project Management and **Programming & Analysis** as well as **Construction & Evaluation** each have 95 questions and NCARB allows you 3 hours and 15 minutes to complete each exam, so you should spend an average of (3x60+15)/80=195/95= 2.05 minutes on each question.

Project Planning & Design as well as **Project Development & Documentation** each have 120 questions and NCARB allows you 4 hours and 15 minutes to complete each exam, so you should spend an average of (4x60+15)/120=255/120= 2.13 minutes on each question.

To simplify this, we suggest you spend about 2 minutes for each question in ALL divisions.

8. **How are ARE scores reported?**
   All ARE scores are reported as Pass or Fail. ARE scores are typically posted within 7 to 10 days. See the instructions in the new ARE guideline at the following link:
   http://www.ncarb.org/ARE/ARE5.aspx

9. **Is there a fixed percentage of candidates who pass the ARE exams?**
   No, there is NOT a fixed percentage of passing or failing. If you meet the minimum competency required to practice as an architect, you pass. The passing scores are the same for all Boards of Architecture.

10. **When can I retake a failed ARE division?**
    You can retake a failed division of the ARE 60 days after the previous attempt. You can only take the same ARE division three (3) times within any 12-month period.

11. **How much time do I need to prepare for each ARE division?**
    Every person is different, but on average you need about 40 to 80 hours to prepare for each ARE division. You need to set a realistic study schedule and stick with it. Make sure you allow time for personal and recreational commitments. If you are working full time, my suggestion is that you allow no less than 2 weeks but NOT more than 2 months to prepare for each ARE division. You should NOT drag out the exam prep process too long and risk losing your momentum.

12. **Which ARE division should I take first?**
    This is a matter of personal preference, and you should make the final decision.

    Some people like to start with the easier divisions and pass them first. This way, they build more confidence as they study and pass each division.

    Other people like to start with the more difficult divisions so that if they fail, they can keep busy studying and taking the other divisions while the clock is ticking. Before they know it, six months has passed and they can reschedule if need be.

13. **ARE exam prep and test-taking tips**
    You can start with Construction & Evaluation (CE) because it gives a limited scope, and you may want to study building regulations and architectural history (especially famous architects and buildings that set the trends at critical turning points) before you take other divisions.

Complete mock exams and practice questions and vignettes, including those provided by NCARB's practice program and this book, to hone your skills.

Form study groups and learn the exam experience of other ARE candidates. The forum at our website is a helpful resource. See the following links:
http://GreenExamEducation.com/
http://GeeForum.com/

Take the ARE exams as soon as you become eligible, since you probably still remember portions of what you learned in architectural school, especially structural and architectural history. Do not make excuses for yourself and put off the exams.

The following test-taking tips may help you:
- Pace yourself properly. You should spend about two minutes for each question on average.
- Read the questions carefully and pay attention to words like *best, could, not, always, never, seldom, may, false, except,* etc.
- For questions that you are not sure of, eliminate the obvious wrong answer and then make an educated guess. Please note that if you do NOT answer the question, you automatically lose the point. If you guess, you at least have a chance to get it right.
- If you have no idea what the correct answer is and cannot eliminate any obvious wrong answers, then do not waste too much time on the question and just guess. Try to use the same guess answer for all of the questions you have no idea about. For example, if you choose "d" as the guess answer, then you should be consistent and use "d" whenever you have no clue. This way, you are likely have a better chance at guessing more answers correctly.
- Mark the difficult questions, answer them, and come back to review them AFTER you finish all MC questions. If you are still not sure, go with your first choice. Your first choice is often the best choice.
- You really need to spend time practicing to become VERY familiar with NCARB's question types. This is because ARE is a timed test, and you do NOT have time to learn about the question types during the test. If you do not know them well, you will NOT be able to finish your solution to the vignette on time.
- The ARE exams test a candidate's competency to provide professional services protecting the <u>health, safety, and welfare</u> of the public. Do NOT waste time on aesthetic or other design elements not required by the program.

ARE exams are difficult, but if you study hard and prepare well, combined with your experience, AXP training, and/or college education, you should be able to pass all divisions and eventually be able to call yourself an architect.

14. Strategies for passing ARE exams on the first try
Passing ARE exams on the first try, like everything else, needs not only hard work, but also great strategy.

- **Find out how much you already know and what you should study**

You goal is NOT to read all the study materials. Your goal is to pass the exam. Flip through the study materials. If you already know the information, skip these parts.

Complete the NCARB sample questions for the ARE exam you are preparing for NOW without ANY studying. See what percentage you get right. If you get 60% right, you should be able to pass the real exam without any studying. If you get 50% right, then you just need 10% more to pass.

This "truth-finding" exam or exercise will also help you to find out what your weakness areas are, and what to focus on.

Look at the same questions again at the end of your exam prep, and check the differences.

Note: We suggest you study the sample questions in the official NCARB Study Guide first, and then other study materials, and then come back to NCARB sample questions again several days before the real ARE exam.

- **Cherish and effectively use your limited time and effort**

    Let me paraphrase a story.
    One time someone had a chance to talk with Napoleon. He said:
    "You are such a great leader and have won so many battles, that you can use one of your soldiers to defeat ten enemy soldiers."

    Napoleon responded:
    "That may be true, but I always try to create opportunities where ten of my soldiers fight one enemy soldier. That is why I have won so many battles."

    Whether this story is true is irrelevant. The important thing that you need to know is **how to concentrate your limited time and effort to achieve your goal. Do NOT spread yourself too thin**. This is a principle many great leaders know and use and is why great leaders can use ordinary people to achieve extraordinary goals.

    Time and effort is the most valuable asset of a candidate. How to cherish and effectively use your limited time and effort is the key to passing any exam.

    If you study very hard and read many books, you are probably wasting your time. You are much better off picking one or two good books, covering the major framework of your exams, and then doing two sets of mock exams to find your weaknesses. You WILL pass if you follow this advice. You may still have minor weakness, but you will have covered your major bases.

- **Do NOT stretch your exam prep process too long**
    If you do this, it will hurt instead of help you. You may forget the information by the time you take the exam.

Spend 40 to 80 hours for each division (a maximum of two months for the most difficult exams if you really need more time) should be enough. Once you decide on taking an exam, put in 100% of your effort and read the RIGHT materials. Allocate your time and effort on the most important materials, and you will pass.

- **Resist the temptation to read too many books and limit your time and effort to read only a few selected books or a few sections of each book in detail**
Having all the books but not reading them, or digesting ALL the information in them will not help you. It is like someone having a garage full of foods, and not eating or eating too much of them. Neither way will help.

    You can only eat three meals a day. Similarly, you can ONLY absorb a certain amount of information during your exam prep. So, focus on the most important stuff.

    Focus on your weaknesses but still read the other info. The key is to understand, digest the materials, and retain the information.

    It is NOT how much you have read, but how much you understand, digest, and retain that counts.

    The key to passing an ARE exam, or any other exam, is to know the scope of the exam, and not to read too many books. Select one or two really good books and focus on them. Actually underline the content and memorize it. For your convenience, I have underlined the fundamental information that I think is very important. You definitely need to memorize all the information that I have underlined.

    You should try to understand the content first, and then memorize the content of the book by reading it multiple times. This is a much better way than relying on "mechanical" memory without understanding.

    When you read the materials, ALWAYS keep the following in mind:

- **Think like an architect.**
For example, when you take the Project Development & Documentation (PDD) exam, focus on what need to know to be able to coordinate your engineer's work, or tell them what to do. You are NOT taking an exam for becoming a structural engineer; you are taking an exam to become an architect.

    This criterion will help you filter out the materials that are irrelevant, and focus on the right information. You will know what to flip through, what to read line by line, and what to read multiple times.

    I have said this one thousand times, and I will say it once more:
    Time and effort is the most valuable asset of a candidate. How to cherish and effectively

use your limited time and effort is the key to passing any exam.

15. ARE exam preparation requires short-term memory

You should understand that ARE Exam Preparation requires **Short-Term Memory**. This is especially true for the MC portion of the exam. You should schedule your time accordingly: in the early stages of your ARE exam Preparation, you should focus on understanding and an **initial** review of the material; in the late stages of your exam preparation, you should focus on memorizing the material as a **final** review.

16. Allocation of your time and scheduling

You should spend about 60% of your effort on the most important and fundamental study materials, about 30% of your effort on mock exams, and the remaining 10% on improving your weakest areas, i.e., reading and reviewing the questions that you answered incorrectly, reinforcing the portions that you have a hard time memorizing, etc.

Do NOT spend too much time looking for obscure ARE information because the NCARB will HAVE to test you on the most **common** architectural knowledge and information. At least 80% to 90% of the exam content will have to be the most common, important and fundamental knowledge. The exam writers can word their questions to be tricky or confusing, but they have to limit themselves to the important content; otherwise, their tests will NOT be legally defensible. At most, 10% of their test content can be obscure information. You only need to answer about 60% of all the questions correctly. So, if you master the common ARE knowledge (applicable to 90% of the questions) and use the guess technique for the remaining 10% of the questions on the obscure ARE content, you will do well and pass the exam.

On the other hand, if you focus on the obscure ARE knowledge, you may answer the entire 10% obscure portion of the exam correctly, but only answer half of the remaining 90% of the common ARE knowledge questions correctly, and you will fail the exam. That is why we have seen many smart people who can answer very difficult ARE questions correctly because they are able to look them up and do quality research. However, they often end up failing ARE exams because they cannot memorize the common ARE knowledge needed on the day of the exam. ARE exams are NOT an open-book exams, and you cannot look up information during the exam.

The **process of memorization** is like **filling a cup with a hole at the bottom**: You need to fill it faster than the water leaks out at the bottom, and you need to constantly fill it; otherwise, it will quickly be empty.

Once you memorize something, your brain has already started the process of forgetting it. It is natural. That is how we have enough space left in our brain to remember the really important things.

It is tough to fight against your brain's natural tendency to forget things. Acknowledging this truth and the fact that you can**not** memorize everything you read, you need to focus your limited time, energy and brainpower on the most important issues.

The biggest danger for most people is that they memorize the information in the early stages of their exam preparation, but forget it before or on the day of the exam and still THINK they remember them.

Most people fail the exam NOT because they cannot answer the few "advanced" questions on the exam, but because they have read the information but can NOT recall it on the day of the exam. They spend too much time preparing for the exam, drag the preparation process on too long, seek too much information, go to too many websites, do too many practice questions and too many mock exams (one or two sets of mock exams can be good for you), and **spread themselves too thin**. They end up **missing the most important information** of the exam, and they will fail.

The ARE Mock Exam series along with the tips and methodology in each of the books will help you find and improvement your weakness areas, MEMORIZE the most important aspects of the test to pass the exam ON THE FIRST TRY.

So, if you have a lot of time to prepare for the ARE exams, you should plan your effort accordingly. You want your ARE knowledge to peak at the time of the exam, not before or after.

For example, if you have two months to prepare for a very difficult ARE exam, you may want to spend the first month focused on reading and understanding all of the study materials you can find as your **initial** review. Also during this first month, you can start memorizing after you understand the materials as long as you know you HAVE to review the materials again later to retain them. If you have memorized something once, it is easier to memorize it again later.

Next, you can spend two weeks focused on memorizing the material. You need to review the material at least three times. You can then spend one week on mock exams. The last week before the exam, focus on retaining your knowledge and reinforcing your weakest areas. Read the mistakes that you have made and think about how to avoid them during the real exam. Set aside a mock exam that you have not taken and take it seven days before test day. This will alert you to your weaknesses and provide direction for the remainder of your studies.

If you have one week to prepare for the exam, you can spend two days reading and understanding the study material, two days repeating and memorizing the material, two days on mock exams, and one day retaining the knowledge and enforcing your weakest areas.

The last one to two weeks before an exam is absolutely critical. You need to have the "do or die" mentality and be ready to study hard to pass the exam on your first try. That is how some people are able to pass an ARE exam with only one week of preparation.

17. Timing of review: the 3016 rule; memorization methods, tips, suggestions, and mnemonics

Another important strategy is to review the material in a timely manner. Some people say that the best time to <u>review</u> material is between <u>30 minutes and 16 hours</u> (the **3016** rule) after you read it for the first time. So, if you review the material right after you read it for the first time, the review may not be helpful.

I have personally found this method extremely beneficial. The best way for me to memorize study materials is to review what I learn during the day again in the evening. This, of course, happens to fall within the timing range mentioned above.

Now that you know the **3016** rule, you may want to schedule your review accordingly. For example, you may want to read <u>new</u> study materials in the morning and afternoon, then after dinner do an <u>initial review</u> of what you learned during the day.

OR

If you are working full time, you can read <u>new</u> study materials in the evening or at night and then get up early the next morning to spend one or two hours on an <u>initial review</u> of what you learned the night before.

The <u>initial</u> review and memorization will make your <u>final</u> review and memorization much easier.

Mnemonics is a very good way for you to memorize facts and data that are otherwise very hard to memorize. It is often <u>arbitrary</u> or <u>illogical</u> but it works.

A good mnemonic can help you remember something for a long time or even a lifetime after reading it just once. Without the mnemonics, you may read the same thing many times and still not be able to memorize it.

There are a few common Mnemonics:
1) <u>**Visual**</u> Mnemonics: Link what you want to memorize to a visual image.
2) <u>**Spatial**</u> Mnemonics: link what you want to memorize to a space, and the order of things in it.
3) <u>**Group**</u> Mnemonics: <u>Break up</u> a difficult piece <u>into</u> several smaller and more <u>manageable groups or sets</u>, and memorize the sets and their order. One example is the grouping of the 10-digit phone number into three groups in the U.S. This makes the number much easier to memorize.
4) <u>**Architectural**</u> Mnemonics: A combination of <u>Visual</u> Mnemonics and <u>Spatial</u> Mnemonics and <u>Group</u> Mnemonics.

Imagine you are walking through a building several times, along the same path. You should be able to remember the order of each room. You can then break up the information that you want to remember and link them to several images, and then imagine you hang the images on walls of various rooms. You should be able to easily recall each group in an orderly manner by imagining you are walking through the building again on the same path, and looking at the images hanging on walls of each room. When you look at the images on the wall, you can easily recall the related information.

You can use your home, office or another building that you are <u>familiar with</u> to build an <u>Architectural</u> Mnemonics to help you to organize the things you need to memorize.

5) **<u>Association</u>** Mnemonics: You can <u>associate</u> what you want to memorize <u>with a sentence</u>, a similarly pronounced word, or a place you are familiar with, etc.
6) **<u>Emotion</u>** Mnemonics: Use emotion to fix an image in your memory.
7) **<u>First Letter</u>** Mnemonics: You can use the <u>first letter</u> of what you want to memorize <u>to construct a sentence or acronym</u>. For example, **"Roy G. Biv"** can be used to memorize the order of the 7 colors of the rainbow, it is composed of the first letter of each primary color.

You can use **Association** Mnemonics and memorize them as <u>all</u> the plumbing fixtures for a typical home, PLUS Urinal.

OR
You can use "Water S K U L" (**<u>First Letter</u>** Mnemonics selected from website below) to memorize them:

<u>Water</u> Closets
<u>S</u>hower
<u>K</u>itchen Sinks
<u>U</u>rinal
<u>L</u>avatory

18. The importance of good and effective study methods

There is a saying: Give a man a fish, feed him for a day. Teach a man to fish, feed him for a lifetime. I think there is some truth to this. Similarly, it is better to teach someone HOW to study than just give him good study materials. In this book, I give you good study materials to save you time, but more importantly, I want to teach you effective study methods so that you can not only study and pass ARE exams, but also so that you will benefit throughout the rest of your life for anything else you need to study or achieve. For example, I give you samples of mnemonics, but I also teach you the more important thing: HOW to make mnemonics.

Often in the same class, all the students study almost the SAME materials, but there are some students that always manage to stay at the top of the class and get good grades on exams. Why? One very important factor is they have good study methods.

Hard work is important, but it needs to be combined with effective study methods. I think people need to work hard AND work SMART to be successful at their work, career, or anything else they are pursuing.

19. The importance of repetition: read this book <u>at least</u> three times

Repetition is one of the most important tips for learning. That is why I have listed it under a separate title. For example, you should treat this book as part of the core study materials for your ARE exams and you need to read this book <u>at least three times</u> to get all of its benefits:

1) The first time you read it, it is new information. You should focus on understanding and digesting the materials, and also do an <u>initial</u> review with the **3016** rule.
2) The second time you read it, focus on reading the parts <u>I</u> have already highlighted AND <u>you</u> have <u>highlighted</u> (the important parts and the weakest parts for you).
3) The third time, focus on <u>memorizing</u> the information.

Remember the analogy of the <u>memorization process</u> as **filling a cup with a hole on the bottom**?
Do NOT stop reading this book until you pass the real exam.

20. The importance of a routine

A routine is very important for studying. You should try to set up a routine that works for you. First, look at how much time you have to prepare for the exam, and then adjust your current routine to include exam preparation. Once you set up the routine, stick with it.

For example, you can spend from 8:00 a.m. to 12:00 noon, and 1:00 p.m. to 5:00 p.m. on studying new materials, and 7:00 p.m. to 10:00 p.m. to do an initial review of what you learned during the daytime. Then, switch your study content to mock exams, memorization and retention when it gets close to the exam date. This way, you have 11 hours for exam preparation everyday. You can probably pass an ARE exam in one week with this method. Just keep repeating it as a way to <u>retain</u> the architectural knowledge.

OR

You can spend 7:00 p.m. to 10:00 p.m. on studying new materials, and 6:00 a.m. to 7:00 a.m. to do an initial review of what you learned the evening before. This way, you have four hours for exam preparation every day. You can probably pass an ARE exam in two weeks with this preparation schedule.

A routine can help you to memorize important information because it makes it easier for you to concentrate and work with your body clock.

Do NOT become panicked and change your routine as the exam date gets closer. It will not help to change your routine and pull all-nighters right before the exam. In fact, if you pull an all-nighter the night before the exam, you may do much worse than you would have done if you kept your routine.

All-nighters or staying up late are not effective. For example, if you break your routine and stay up one-hour late, you will feel tired the next day. You may even have to sleep a few more hours the next day, adversely affecting your study regimen.

21. The importance of short, frequent breaks and physical exercise

Short, frequent breaks and physical exercise are VERY important for you, especially when you are spending a lot of time studying. They help relax your body and mind, making it much easier for you to concentrate when you study. They make you more efficient.

Take a five-minute break, such as a walk, at least once every one to two hours. Do at least 30 minutes of physical exercise every day.

If you feel tired and cannot concentrate, stop, go outside, and take a five-minute walk. You will feel much better when you come back.

You need your body and brain to work well to be effective with your studying. Take good care of them. You need them to be well-maintained and in excellent condition. You need to be able to count on them when you need them.

If you do not feel like studying, maybe you can start a little bit on your studies. Just casually read a few pages. Very soon, your body and mind will warm up and you will get into study mode.

Find a room where you will NOT be disturbed when you study. A good study environment is essential for concentration.

22. A strong vision and a clear goal

You need to have a strong vision and a clear goal: to <u>master</u> the architectural knowledge and <u>become an architect in the shortest time</u>. This is your number one priority. You need to master the architectural knowledge BEFORE you do sample questions or mock exams, except "truth-finding" exam or exercise at the very beginning of your exam prep. It will make the process much easier. Everything we discuss is to help you achieve this goal.

As I have mentioned on many occasions, and I say it one more time here because it is so important:

It is how much architectural knowledge and information you can <u>understand, digest, memorize</u>, and firmly retain that matters, not how many books you read or how many sample tests you have taken. The books and sample tests will NOT help you if you cannot understand, digest, memorize, and retain the important information for the ARE exam.

Cherish your limited <u>time and effort</u> and focus on the most <u>important</u> information.

23. English system (English or inch-pound units) vs. metric system (SI units)

This book is based on the English system or English units. The English or inch-pound units are based on the module used in the U.S. Effective July 2013, the ARE includes measurements in inch-pound units only. Metric system (SI units) is no longer used.

24. Codes and standards used in this book

We use the following codes and standards:

American Institute of Architects, Contract Documents, Washington, DC; ADA Standards for Accessible Design, ADA; Various International Codes by ICC. See Appendixes for more information.

25. Where can I find study materials on architectural history?
Every ARE exam may have a few questions related to architectural history. The following are some helpful links to FREE study materials on the topic:
http://issuu.com/motimar/docs/history_synopsis?viewMode=magazine

Chapter Two

Project Planning & Design (PPD) Division

A. General Information

1. Exam content

The PPD division of the ARE has 120 questions which cover five different areas.

Sections	Target Percentage	Expected Number of Items
Section 1: Environmental Conditions & Context	10-16%	12-20
Section 2: Codes & Regulations	16-22%	20-27
Section 3: Building Systems, Materials, & Assemblies	19-25%	22-30
Section 4: Project Integration of Program & Systems	32-38%	38-46
Section 5: Project Costs & Budgeting	8-14%	9-17

The exam content can be further broken down as follows:

Section 1: Environmental Conditions & Context (10-16%)
- Determine location of building and site improvements based on site analysis (A/E)
- Determine sustainable principles to apply to design (A/E)
- Determine impact of neighborhood context on the project design (U/A)

Section 2: Codes & Regulations (16-22%)
- Apply zoning and environmental regulations to site and building design (U/A)
- Apply building codes to building design (U/A)
- Integrate multiple codes to a project design (A/E)

Section 3: Building Systems, Materials, & Assemblies (19-25%)
- Determine mechanical, electrical, and plumbing (MEP) systems (A/E)
- Determine structural systems (A/E)
- Determine special systems such as acoustics, communications, lighting, security, conveying, and fire suppression (A/E)
- Determine materials and assemblies to meet programmatic, budgetary, and regulatory requirements (A/E)

Section 4: Project Integration of Program & Systems (32-38%)
- Determine building configuration (A/E)
- Integrate building systems in the project design (A/E)
- Integrate program requirements into a project design (A/E)
- Integrate environmental and contextual conditions in the project design (A/E)

Section 5: Project Costs & Budgeting (8-14%)
- Evaluate design alternatives based on the program (A/E)
- Perform cost evaluation (A/E)
- Evaluate cost considerations during the design process (A/E)

2. **Official exam guide and reference index for the Project Planning & Design (PPD) division**

   NCARB published the exam guides for all ARE 5.0 division together as *ARE 5.0 Handbook*.

   You need to read the official exam guide for the PPD division at least three times and become very familiar with it. The real exam is VERY similar to the sample questions in the handbook.

   You can download the official *ARE 5.0 Handbook* at the following link:
   http://blog.ncarb.org/2016/November/ARE5-Study-Materials.aspx

   Note: We suggest you study the official ARE 5.0 Handbook first, and then other study materials, and then come back to Handbook again several days before the real ARE exam.

B. **The Most Important Documents/Publications for PPD Division of the ARE Exam**

1. **Official NCARB list of formulas and references for the Project Planning & Design (PPD) division with our comments and suggestions**
   You can find the NCARB list of references for the PPD division in the Appendixes of this book and the *ARE 5.0 Handbook*.

   The formulas will be available during the real exam. You should read through them a few times before the exam to become familiar with the content. This will save you a lot of time during the real exam, and will help you solve structural calculations and other problems.

   Note:
   *While many of the MC questions in the real PPD ARE exam **focus on structural design concepts**, there are **some questions requiring calculations**. Therefore, it is absolutely necessary and critical for you to be very familiar with some of the basic and important*

equations, and to memorize them if possible. We have incorporated some of the most important equations into our PPD mock exam.

In the ARE exams, it may be a good idea to skip any calculation question that requires over 30 seconds of your time; just pick a guess answer, mark it, and come back to calculate it at the end. This way, you have more time to read and answer other easier questions correctly.

A calculation question that takes 20 minutes to answer will gain the same number of points as a simple question that ONLY takes 2 minutes.

If you spend 20 minutes on a calculation question earlier, you risk losing the time to read and answer ten other easier questions, which could result in a loss of ten points instead of one.

The NCARB list of references includes the following:

Publications
Architectural Graphic Standards
The American Institute of Architects
John Wiley & Sons, latest edition

Building Codes Illustrated: A Guide to Understanding the 2012 International Building Code
Francis D.K. Ching and Steven R. Winkel, FAIA, PE
John Wiley & Sons, 2012

Building Structures
James Ambrose and Patrick Tripeny
John Wiley & Sons, 3rd edition, 2012

Fundamentals of Building Construction: Materials and Methods
Edward Allen and Joseph Iano
John Wiley & Sons, latest edition

Mechanical & Electrical Equipment for Buildings
Walter T. Grondzik, Alison G. Kwok, Benjamin Stein, and John S. Reynolds, Editors
John Wiley & Sons, latest edition

Codes
2010 ADA Standards for Accessible Design
U.S. Department of Justice, 2010

2012 International Building Code (IBC). International Code Council, Inc. Country Club Hills, Illinois, 2011

Focus on **Chapter 16**, particularly the sections on earthquakes and wind. Read them a few times, and have a general idea of the concepts. Do not force yourself to memorize all the details.

Read the following pages and become familiar with the Uniform and Concentrated Loads IBC table 1607.1:
pg. 285 & 286.
OR
You can read the Uniform and Concentrated Loads IBC Table 1607.1 for FREE at the following link:
http://publiccodes.cyberregs.com/icod/ibc/2006f2/icod_ibc_2006f2_16_sec001.htm

Note:
- *The latest version of IBC is the 2015 version. For your convenience, we provide a link to the free online version of the IBC. The basic IBC content stays the same for many years. If the codes are updated by NCARB in the future, you just need to go to the root directory and find the latest version of the codes:*
  http://codes.iccsafe.org

- *You need to spend a large percentage (at least 20%) of your prep time on IBC Chapter 16 and become familiar with it. You do not need to force yourself to memorize the numbers and all the detail. Just reading it a few times and becoming familiar with the information should be adequate.*

AIA Contract Documents
None of the standard list of AIA Contract Documents related to the ARE have specific content covered in the Project Planning & Design division.

The following are some extra study materials that are useful if you have some additional time and want to learn more. If you are tight on time, you can simply look through them and focus on the sections that cover your weaknesses:

2. *Manual of Steel Construction: Allowable Stress Design*; 9th Edition.
   American Institute of Steel Construction, Inc. Chicago, Illinois, 1989

Look through the following pages and become familiar with the structural shapes, designations, dimensions, and properties:
pg. 1-9 thru 1-16, *pg.* 1-18 thru 1-32, *pg.* 1-40 thru 1-41, *pg.* 1-46 thru 1-52, *pg.* 1-94 thru 1-103.

Make sure you understand what the designations stand for. For example, on pages 1-10 and 1-11, **W 40 x 298** means a W shape steel with a nominal depth of 40" (the actual depth is 39.69" per the Table on page 1-10), and a nominal weight of 298 lb. per ft. Use the diagram and Tables on pages 1-10 and 1-11 to look up the other detailed properties of a **W**

40 x 298. You do NOT need to remember any of these properties. You just need to know how to look them up and what they mean.

The most important information for an architect is the overall dimension of a structural member so that you can coordinate and make sure you have enough space to accommodate it. For example, you may need to find out if it will fit inside a wall or interstitial space.

You can also use the size of a structural member, the mechanical duct size, and the clearance space needed for a light fixture and/or fire sprinkler line to determine the interstitial height between floors.

Look through the following pages and become familiar with the beam nomenclature, diagrams, and formulas:
pg. 2-293 & 2-294, pg. 2-296, pg. 2-297, pg. 2-298, pg. 2-301, pg. 2-304, pg. 2-305.

Look through the following pages and become familiar with bolts, threaded parts, and rivet tensions:
pg. 4-3 & 4-5.

3. *Steel Construction Manual*; 13th Edition. American Institute of Steel Construction, Inc. Chicago, Illinois, 2005
   Read the following pages and become familiar with the round HSS dimensions and properties:
   pg. 1-94 thru 1-98.

4. **FREE information on truss and beam diagrams** can be found at the following link:
   http://ocw.mit.edu/ans7870/4/4.463/f04/module/Start.html

5. **The FREE PDF file of FEMA publication number 454 (FEMA454),** *Designing for Earthquakes: A Manual for Architects*, is available at the following link:
   http://www.fema.gov/library/viewRecord.do?id=2418

   Note:
   *You need to spend a large percentage (at least 30%) of your prep time on this **FEMA454** PDF file. Focus on **Chapters 4, 5, 8, and 9**. You may have **many real ARE questions** based on these chapters. This PDF book has many diagrams and photos and helps you understand what happened in buildings that failed during an earthquake.*

6. **The FREE PDF file of *Wind Design Made Simple*** by ICC TRI-Chapter Uniform Code Committee is available at the following link:
   http://www.calbo.org/Documents/SimplifiedWindHandout.pdf

   Note:
   You need to become familiar with this file. You do not need to force yourself to memorize the numbers and all the details, just reading it a few times and becoming familiar with the information should be adequate.

7. Arnold, Christopher. ***Building at Risk***, is available for FREE at the AIA website: http://www.aia.org/aiaucmp/groups/aia/documents/pdf/aiap016810.pdf

8. **Construction Specifications Institute (CSI) MasterFormat & *Building Construction***
Become familiar with the new 6-digit CSI Construction Specifications Institute (CSI) MasterFormat as there may be a few questions based on this publication. Make sure you know which items/materials belong to which CSI MasterFormat specification section, and memorize the major section names and related numbers. For example, Division 9 is Finishes, and Division 5 is Metal, etc. Another one of my books, *Building Construction*, has detailed discussions on CSI MasterFormat specification sections.

Mnemonics for the 2004 CSI MasterFormat

The following is a good mnemonic, which relates to the 2004 CSI MasterFormat division names. Bold font signals the gaps in the numbering sequence.

This tool can save you lots of time: if you can remember the four sentences below, you can easily memorize the order of the 2004 CSI MasterFormat divisions. The number sequencing is a bit more difficult, but can be mastered if you remember the five bold words and numbers that are not sequential. Memorizing this material will not only help you in several divisions of the ARE, but also in real architectural practice

Mnemonics (pay attention to the underlined letters):
Good students can memorize material when teachers order.
F students earn F's simply 'cause **forgetting** principles have **an** effect. (21 and 25)
C students **end** everyday understanding things without memorizing. (31)
Please make professional pollution prevention inventions **everyday**. (40 and 48)

```
 1-Good...................... General Requirements
 2-Students.................. (Site) now Existing Conditions
 3-Can........................ Concrete
 4-Memorize................. Masonry
 5-Material .................. Metals
 6-When....................... Woods and Plastics
 7-Teachers.................. Thermal and Moisture
 8-Order...................... Openings

 9-F............................. Finishes
10-Students.................. Specialties
11-Earn....................... Equipment
12-F's......................... Furnishings
13-Simply................... Special Construction
14-'Cause.................... Conveying
21-Forgetting .............. Fire
22-Principles............... Plumbing
```

23-Have............................... HVAC
25-An......................................Automation
26-Effect............................... Electric

27-C.. Communication
28-Students........................... Safety & Security
31-End.................................... Earthwork
32-Everyday..........................Exterior
33-Understanding ....Utilities
34-Things.............................. Transportation
35-Without Memorizing........ Waterways and Marine

40-Please...............................Process Integration
41-Make................................ Material Processing and Handling Equipment
42-Professional..................... Process Heating, Cooling, and Drying Equipment
43-Pollution.......................... Process Gas and Liquid Handling, Purification and Storage Equipment
44-Prevention........................Pollution Control Equipment
45-Inventions........................ Industry-Specific Manufacturing Equipment
48-Everyday..........................Electrical Power Generation

Note:
There are 49 CSI divisions. The "missing" divisions are those "reserved for future expansion" by CSI. They are intentionally omitted from the list.

Chapter Three

ARE Mock Exam for
Project Planning & Design (PPD) Division

A. **Multiple-Choice (MC)**

1. Per the *International Building Code (IBC)*, which of the following is correct? **Check the two that apply.**
    a. *IBC* allows live load reduction in most cases.
    b. *IBC* allows live load reduction in a few cases.
    c. *IBC* does not allow live load reduction for public assembly occupancy with a live load equal to or less than 100 psf.
    d. *IBC* does not allow live load reduction for a live load equal to or more than 100 psf.

2. The load of an automobile moving in a parking garage is a
    a. dynamic load
    b. impact load
    c. dead load
    d. vertical load

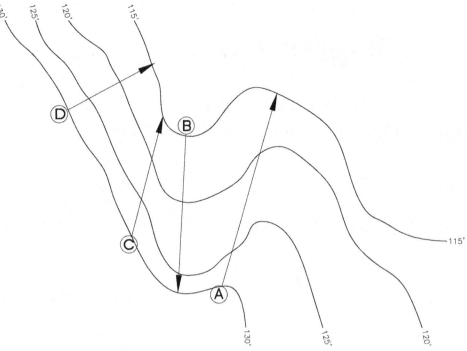

Figure 3.1   Direction of water flow

3. With regard to figure 3.1, which of the following is the direction of water flow?
   a. A
   b. B
   c. C
   d. D

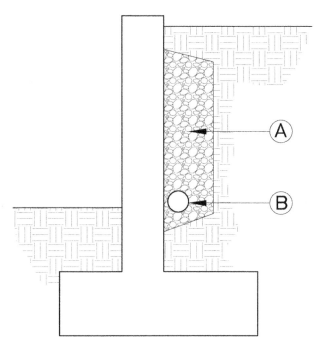

Figure 3.2   Retaining wall section

4. Which of the following is "A" as shown on figure 3.2?
   a. Sand
   b. Soil
   c. Sand and gravels
   d. Gravel

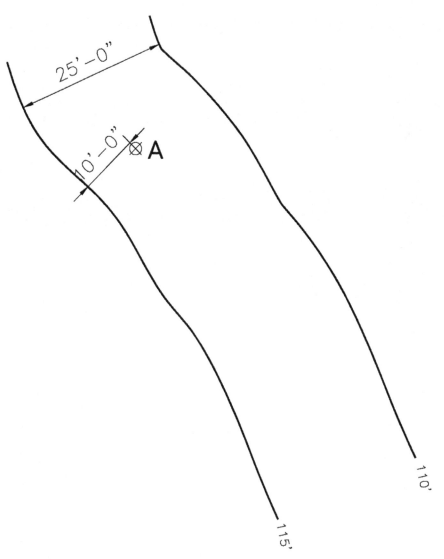

Figure 3.3   Spot elevations

5. Assuming a constant slope between the two contour lines shown, what is the elevation of point A?
   a. 112'
   b. 112.5'
   c. 113'
   d. 113.5'
   e. 114'

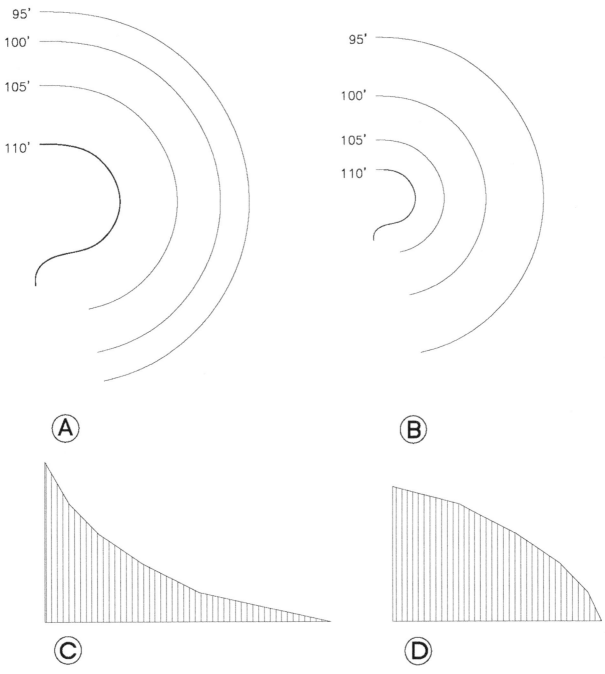

Figure 3.4   Types of slopes

6. Refer to figure 3.4, which of the following shows a convex slope? **Check the two that apply.**
   a. A
   b. B
   c. C
   d. D

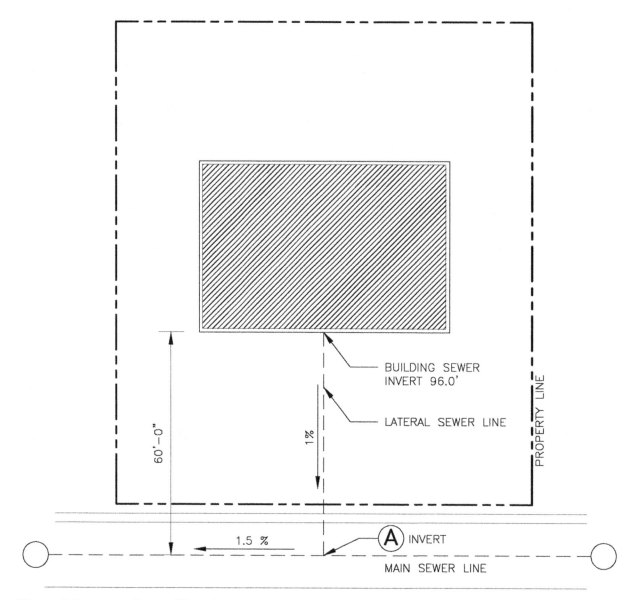

Figure 3.5　　　Invert Elevation

7. If the slope for the lateral sewer line is 1% and the slope for the main sewer line is 1.5%, what is the invert elevation at point A as shown on figure 3.5?
   a. 95.1'
   b. 95.3'
   c. 95.4'
   d. 96.7'

Figure 3.6   Lock image

8. What kind of lock is shown on the previous image?
   a. Mortise lock
   b. Unit lock
   c. Integral lock
   d. Cylinder lock

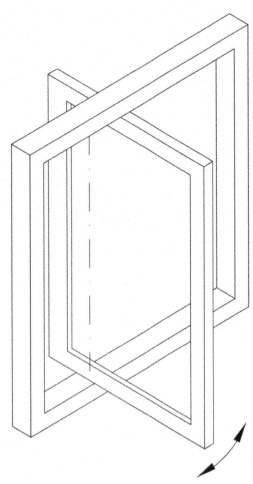

Figure 3.7   Window type

9. What kind of window is shown on the previous image?
   a. Sliding
   b. Pivoting
   c. Casement
   d. Fixed

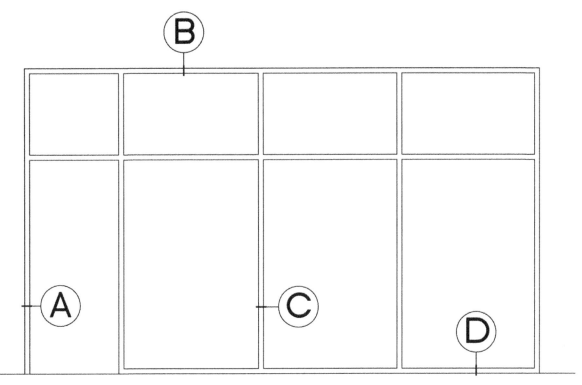

Figure 3.8   Storefront elevation

10. What letter in the previous figure indicates a mullion?
    a. A
    b. B
    c. C
    d. D

11. In a building that uses Insulated Concrete Forms (ICF) system, the forms are left in place permanently for the following reasons: (**Check the four that apply**)
    a. Thermal insulation
    b. Acoustic insulation
    c. Space to run plumbing pipes and electrical conduits.
    d. Aesthetic effect
    e. Construction cost
    f. Backing for gypsum boards, stucco, and brick

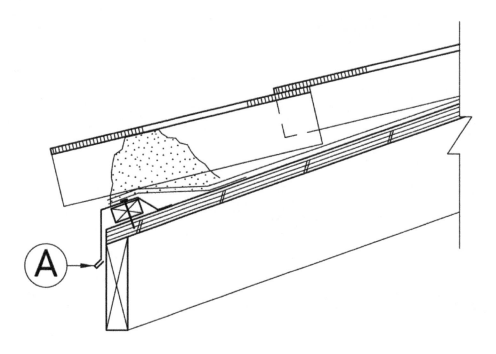

Figure 3.9   Roof detail

12. What is noted as letter A on the previous image?
    a. Flashing
    b. Drip edge
    c. Coping
    d. Siding

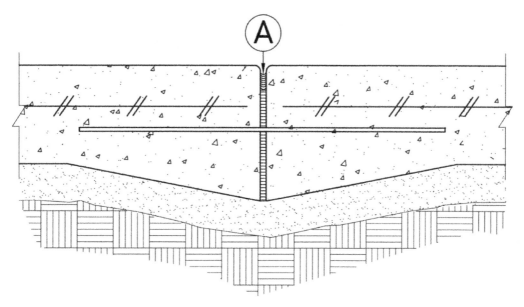

Figure 3.10   Concrete detail

13. What is noted as letter A on the previous image?
    a. Construction joint
    b. Control joint
    c. Concrete separation
    d. Concrete mark

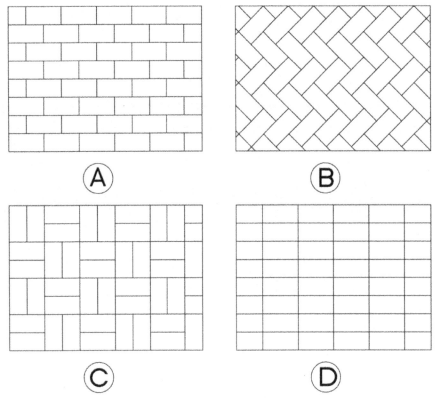

Figure 3.11   Brick pattern

14. Which pattern/letter on the previous image is stack bond?
    a. Pattern A
    b. Pattern B
    c. Pattern C
    d. Pattern D

15. According to *International Building Code* (IBC), interior adhered masonry veneers shall have a maximum weight of _____ psf (or _____ kg/ m²).

16. Per IBC, which of the following is correct regarding the panic hardware at the required exits?
    a. The actuating portion of the releasing device shall extend at least one-quarter of the door leaf width.
    b. The actuating portion of the releasing device shall extend at least one-half of the door leaf width.
    c. The actuating portion of the releasing device shall extend at least three-quarter of the door leaf width.
    d. The actuating portion of the releasing device shall extend the entire door leaf width.

17. Per 2902.1 of IBC, the minimum number of required water closets for Mercantile Occupancy Group is 1 per 500 occupants. An architect is designing a retail store with 32,000 square feet of gross area. All the store areas are on the ground level, and 5% of the areas are storage. The remaining areas are all retail spaces. Based on the following table, the minimum number of water closets required for Women's Restroom for this store is __ _____.

Table 3.1   Maximum Floor Area Allowances per Occupant

Mercantile	
Areas on other floors	60 gross
Basement and grade floor areas	30 gross
Storage, stock, shipping areas	300 gross

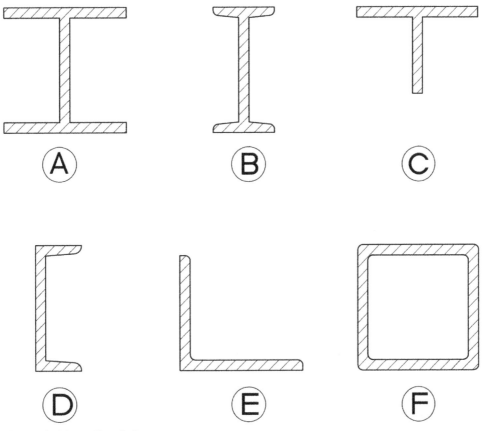

Figure 3.12   Steel shapes

18. Which of the previous images shows a W shape steel?
    a. A
    b. B
    c. C
    d. D
    e. E
    f. F

19. Per *International Mechanical Code* (IMC), if the height of a roof access ladder is over _____ feet, an intermediate landing is required.

20. The nominal size of a standard brick in the US is _____, and the nominal size of a standard concrete masonry unit (CMU) in the US is _____. The actual size is usually about _____ smaller to allow for mortar joints.

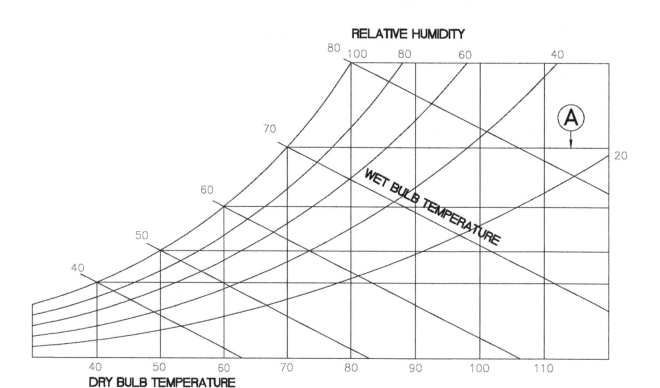

Figure 3.13   **Chart** (The temperatures are shown in °F, and the relative humidity is shown in percentages.)

21. The line labeled as "A" in figure 3.13 represents_____.

22. An architect is designing an office with 1,500 s.f. (139.35 s.m.) of gross area. The owner's design criteria require 10,000 BTU per 300 s.f. of gross area (293.071 W per 27.87 s.m.). If the market price of a central air-conditioning system is $1,000 per ton and the system is only available in an integer ton, what is the minimum cost of a system for the office that will meet the owner's criteria?
    _____ dollars

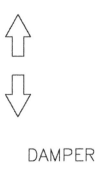

DAMPER

Ⓣ THERMOSTAT

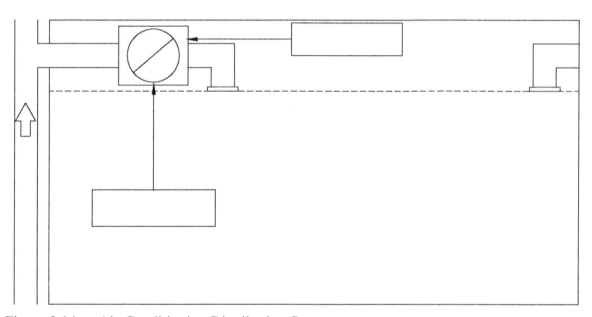

Figure 3.14   Air-Conditioning Distribution System

23. Drag the labels and symbols from the area on the top onto the schematic layout of the recommended system below. Not all items will be used. What type of air-conditioning distribution system is shown in figure 3.14?
   a. a single zone system
   b. a variable flow system
   c. a double duct system
   d. a terminal reheat system

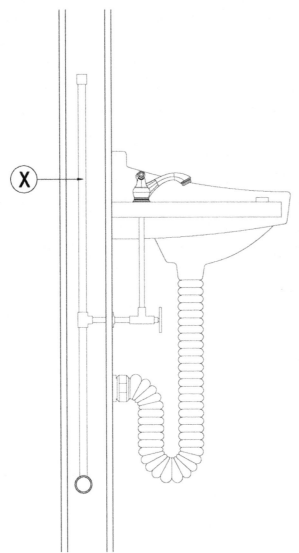

Figure 3.15   Plumbing Drawing

24. What is the term for X in figure 3.15?
    a. exhaust vent
    b. manufactured shock absorber
    c. capped air chamber
    d. vent pipe

25. For a roof space accessible by the public or building tenants, what is the minimum height of a vent pipe extension?
    a. 6" above the roof
    b. 8" above the roof
    c. 6'-8" above the roof
    d. 7'-0" above the roof

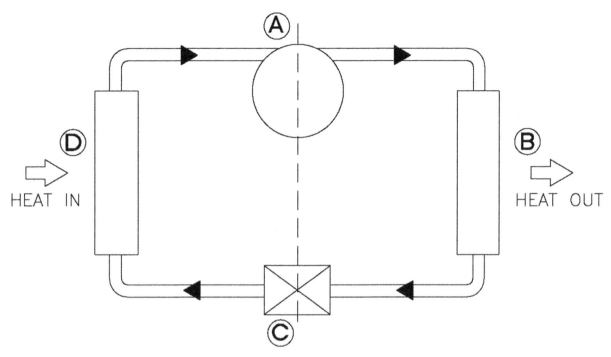

Figure 3.16   Refrigeration Flow Diagram

26. In the refrigeration flow diagram of figure 3.16, where is expansion valve located?
    a. A
    b. B
    c. C
    d. D

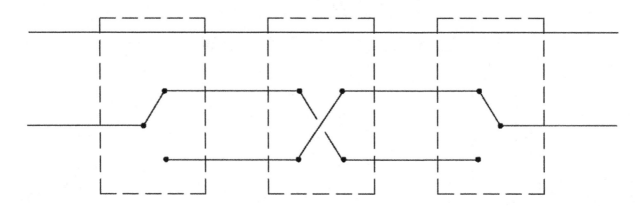

Figure 3.17   Switch Wiring Diagram

27. Which of the following devices control lighting from three different locations as indicated in figure 3.17? **Check the two that apply.**
   a. two single-pole, single-throw switches
   b. two single-pole, double-throw switches
   c. one four-way switch
   d. one three-pole, double-throw switch

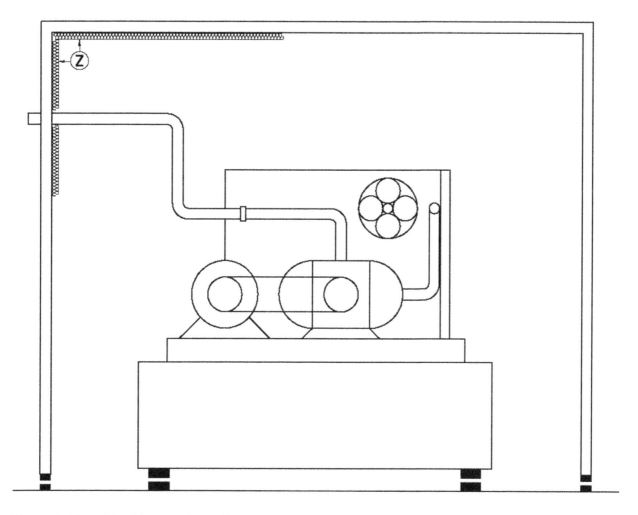

Figure 3.18   Machines and Appliances

28. What is the term for Z in figure 3.18?
   a. resilient hanger
   b. acoustical lining
   c. paver pedestal
   d. vibration isolator

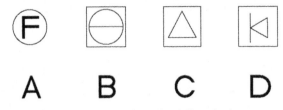

Figure 3.19   Electrical Symbols

29. Which symbol in figure 3.19 is a floor special purpose outlet?
   a. A
   b. B
   c. C
   d. D

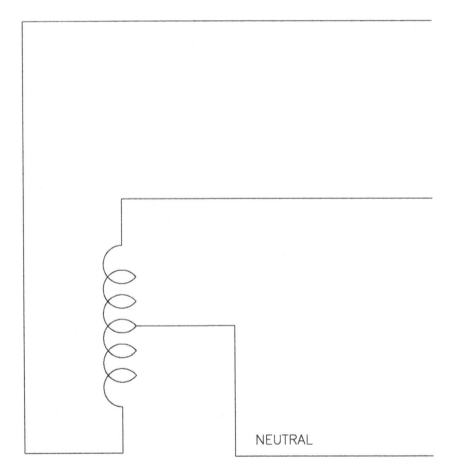

Figure 3.20   Wiring Diagram

30. What is the term for the wiring system in figure 3.20?
   a. single-phase, 3-wire service
   b. double-phase, 3-wire service
   c. three-phase, 3-wire service
   d. none of the above

31. An architect needs to calculate the U-value (overall thermal transmittance) for a wall assembly. Based on the following data, what is the U-value for the wall?

Component	R-value
Outside air layer	0.17
3/4" Cement plaster, sand aggregate	0.15
1/2" Plywood	0.62
Nominal 6" batt fiberglass	19.00
Gypsum board	0.45
Inside air layer	0.68

a. approximately 0.02
b. approximately 0.05
c. approximately 0.08
d. approximately 0.10

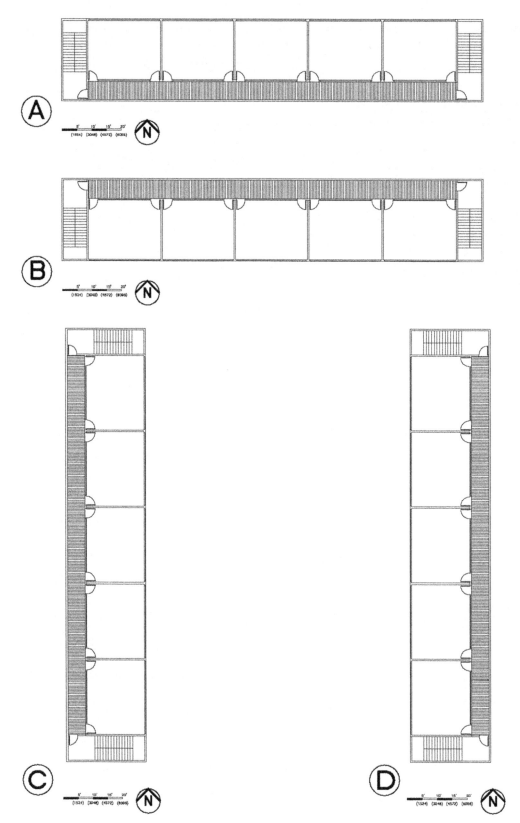

Figure 3.21　　　Orientation of a Classroom Building

32. An architect is working on a classroom building in the Northern Hemisphere. The teachers and students want the classroom to be energy efficient, and have the greatest potential to view outside with minimum window blind use. Which of the options in figure 3.21 is the best orientation for the classroom building to meet the demand of the teachers and students?
    a. A
    b. B
    c. C
    d. D

33. Figure 3.22 shows a truss. Which of the following shows the reactions at A and B correctly?

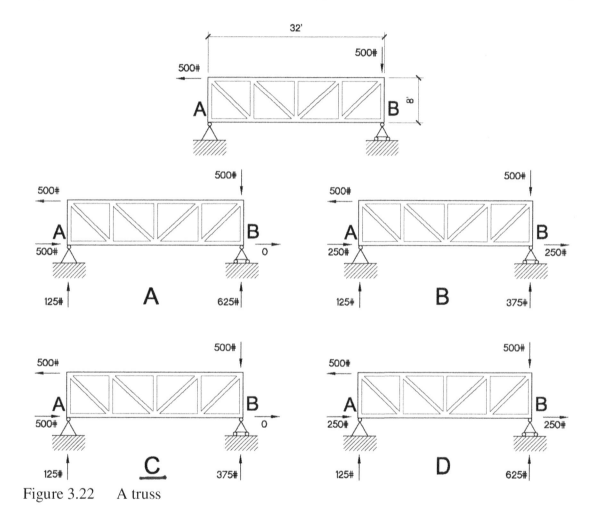

Figure 3.22   A truss

34. Figure 3.23 shows a building's floor plan. The total wind load in the east-west direction
    a. is greater than in the north-south direction
    b. is the same as in the north-south direction
    c. is smaller than in the north-south direction
    d. can be more or less than in the north-south direction, depending on the gust factor

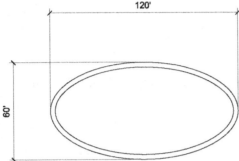

Figure 3.23　A building's floor plan

35. Figure 3.24 shows beam to column connections. Which of the following is part of a moment-resisting frame?
    a. I
    b. II
    c. I and II
    d. neither I or II

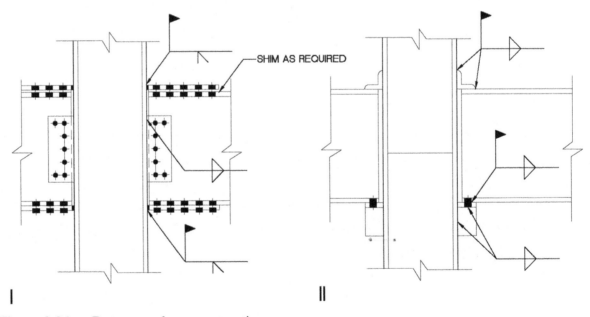

Figure 3.24　Beam to column connections

36. Figure 3.25 shows a moment-resisting frame. Which of the following statements is correct? **Check the two that apply.**
    a. The bottom of the column is free to rotate.
    b. The bottom of the column is fixed against rotation.
    c. The top of the column is free to rotate.
    d. The top of the column is fixed against rotation.

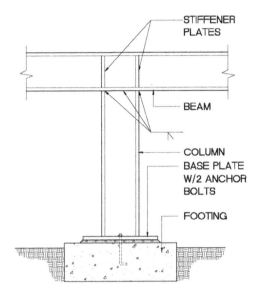

Figure 3.25   Diagram for a frame

37. Which of the diagrams in figure 3.26 show the distribution of lateral forces used in seismic design?

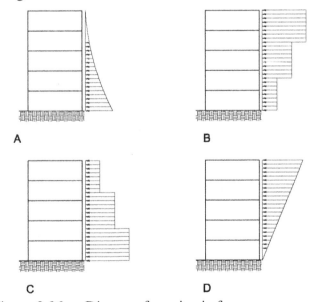

Figure 3.26   Diagram for seismic forces

38. Which of the diagrams in figure 3.27 shows the deflected shape of a rigid frame with a rigid base for the loading shown?

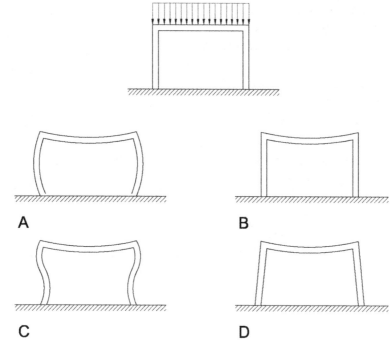

Figure 3.27 Diagram for a rigid frame

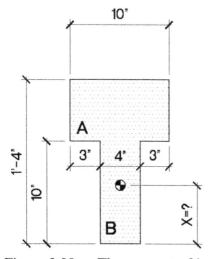

Figure 3.28 The moment of inertia for a composite section

39. The moment of inertia for the composite section as shown in figure 3.28 is _____ _____ ft-lbf.

40. Open-web steel joists spaced 4 ft on center span 36 ft. The dead load = 30 psf (including the weight of the joists), live load = 50 psf, and maximum allowable deflection = 1/360 of the span. Based on table 3.2, what is the most economical section to use?

Table 3.2   Partial Load Table for Open-Web Steel Joists, K-Series
(Based on a maximum allowable tensile stress of 50,000 psi; loads in pounds per linear foot)

Joist Designation	20K9	20K10	22K7	22K9
Depth (in)	20	20	22	22
Approx. Wt. (lbs/ft)	10.8	12.2	9.7	11.3
Span (ft) ↓				
35	329	390	303	364
	179	210	185	219
36	311	369	286	344
	164	193	169	201

a. 20K9
b. 20K10
c. 22K7
d. 22K9

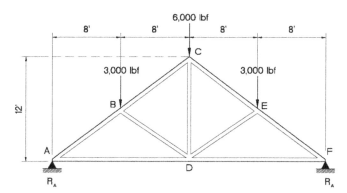

Figure 3.29   A simple truss

41. Figure 3.29 shows a simple truss. Using the method of joints, the force in member AB (the member between joints A and B) is
   a. 10,000 lbf
   b. 11,000 lbf
   c. 12,000 lbf
   d. 13,000 lbf

42. In regard to wind load design, the categories B, C, and D are based on
    a. surface roughness
    b. geography
    c. distance to the ground surface
    d. wind strength

43. Which of the following in figure 3.30 is the correct diagram of wind pressure for a 12-foot high building with a 30-foot square plan?
    a. A
    b. B
    c. C
    d. D

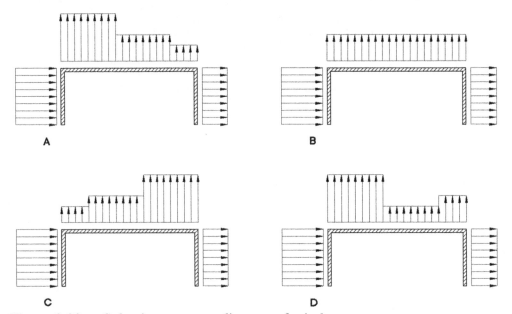

Figure 3.30   Selecting a correct diagram of wind pressure

44. Which of the following regarding the Seismic Design Category (SDC) of a structure is true? **Check the two that apply.**
    a. Buildings in SDC A need to meet more restrictive earthquake requirements than buildings in SDC C.
    b. Buildings in SDC C need to meet more restrictive earthquake requirements than buildings in SDC A.
    c. SDC includes categories A through D.
    d. SDC includes categories A through E.

45. Which of the following is graywater? **Check the two that apply.**
    a. rainwater
    b. water from dishwashers or kitchen sinks
    c. water from laundry tubs and clothes washers
    d. water from toilets

46. Underground water lines should be placed
    a. above the frost line
    b. with the centerline align with the frost line
    c. below the frost line
    d. none of the above

47. The solar altitude is largest in the Northern Hemisphere on the day of the
    a. vernal equinox
    b. summer solstice
    c. autumnal equinox
    d. winter solstice

48. A developer wants to build an elementary school in an area zoned for residential use. Which of the following applications should she submit to the city?
    a. Conditional Use Permit
    b. Non-conforming Use Permit
    c. Incentive Zoning Permit
    d. Ordinance Variance Permit

49. Which of the following statements are correct? **Check the two that apply.**
    a. In a cold climate, an architect should use materials with low albedo and low conductivity for ground surfaces.
    b. In a cold climate, an architect should use materials with low albedo and high conductivity for ground surfaces.
    c. In a tropical climate, an architect should use materials with low albedo and low conductivity for ground surfaces.
    d. In a tropical climate, an architect should use materials with high albedo and high conductivity for ground surfaces.

50. Per the U.S. Environmental Protection Agency (EPA), using a _____-based approach to wetland protection ensures that the whole system, including land, air, and water resources, is protected.

51. Per EPA, which of the following are general categories of wetlands found in the U.S.? **Check the four that apply.**
    a. Bogs
    b. Fens
    c. Lakes
    d. Marshes
    e. Reservoirs
    f. Swamps

52. Per EPA, which of the following is the leading cause of species extinction?
    a. Human activities
    b. Pollution
    c. Habitat degradation
    d. Excessive hunting

53. An architect is working on a residential project. Which of the following are correct statements? **Check the two that apply.**
    a. The architect should review the building codes for FAR information.
    b. The architect should contact the Planning Department to obtain information on the existing easement and set back requirements for the front yard, the side yard and the back yard.
    c. The architect should contact the Planning Department to obtain site coverage ratio.
    d. The architect should contact the Planning Department to obtain zoning information for the site.

54. Which of the following are not effective ways of controlling noise between two adjacent apartment units? **Check the two that apply.**
    a. flexible boots
    b. resilient hangers
    c. acoustical lining outside of the HVAC ducts
    d. acoustical lining inside of the HVAC ducts
    e. using stainless steel ducts

55. Which of the following statements are true? **Check the two that apply.**
    a. Duct silencers and baffles are normally placed inside the HVAC ducts.
    b. Duct silencers and baffles are normally placed outside the HVAC ducts.
    c. Duct silencers and baffles are useful to reduce fan noise but cause considerable pressure drop.
    d. Duct silencers and baffles are not useful to reduce fan noise and cause considerable pressure drop.

56. A tenant wants to install 2'x4' recessed light fixtures in an office. The floor area for the office is 1,200 s.f. Each fixture has four lamps. The tenant wants to have a lighting level of about 50 footcandles (fc) at desk level (3'-0" AFF). The luminaires can provide about 2,800 lumens per lamp at desk level. The coefficient of utilization (CU) is 60% when the fixtures are installed at 10'-0" above finish floor (AFF). Light loss factor (LLF) is 0.65. How many 2'x4' recessed light fixtures does the tenant need for this office?
    a. 12
    b. 14
    c. 16
    d. 18

57. For the same office and criteria in question 36, if the tenant uses 2-lamp 1'x4' fixtures instead of the 2'x4' recessed light fixtures, and these luminaires can provide about 3,000 lumens per lamp at desk level, how many 1'x4' light fixtures does the tenant need for this office?
    a. 26
    b. 28
    c. 30
    d. 32

58. For the same office and criteria in question 36, if utilities cost $5 per year for every 1,000 lumens, how much is the tenant utility cost for this office per month?
    a. $15
    b. $20
    c. $25
    d. $30

59. For the same office in question 36, if the cost of installation, labor, and materials is $300 for each 2'x4' recessed light fixture, how much is the tenant cost for installing all the lights in this office?
    a. $3,600
    b. $4,200
    c. $4,800
    d. $5,400

60. An architect is working on a shopping center. One of the tenants requires 1200A, 277/480 V, 3-phase, 4-wire electrical service. The architect wants to verify if the electrical engineer has provided the requested service on the plans. Where can the architect find this information? **Check the three that apply.**
    a. electrical lighting plans
    b. single-line diagram
    c. panel schedules
    d. low-voltage plans
    e. power plans

61. In the design development phase of a design-bid-build project, an architect has received the mechanical roof plan and HVAC equipment schedules. Which of the project team members should the architect forward this information to for coordination? **Check the three that apply.**
    a. plumbing engineer
    b. electrical engineer
    c. structural engineer
    d. contractor
    e. civil engineer
    f. fire protection engineer

62. A credit card company has purchased a 15-year-old concrete tilt-up warehouse building. The company wants to remodel the warehouse as an office building. Additional HVAC equipment is necessary to provide comfort for building occupants. There is no as-built information for the building. What is the most cost-effective way to achieve this goal?
    a. Add additional columns and beams to support the additional weight of the new HVAC equipment.
    b. Do an X-ray of the building structure to find out if it can support the additional weight of the new HVAC equipment.
    c. Place the additional HVAC equipment on a concrete pad on grade outside of the building and install new ducts to the new HVAC equipment.
    d. Place the new HVAC equipment right on top of the existing structural columns.

63. A fan coil unit (FCU) is
    a. a simple device consisting of a heating or cooling coil and fan
    b. a fire-safety system
    c. an evaporative cooling system
    d. a hot-water recirculation system

64. Which of the following statements are true? **Check the two that apply.**
    a. A dry pipe fire sprinkler system is one in which the pipes are filled with pressurized air, rather than water.
    b. A dry pipe fire sprinkler system is one in which the pipes are filled with pressurized Halon, rather than water.
    c. A dry pipe fire sprinkler system is lighter and less expensive to install than a wet-pipe sprinkler system
    d. A dry pipe fire sprinkler system will not freeze in unheated spaces.
    e. A dry pipe fire sprinkler system has fewer valves and fittings to maintain.

65. Which of the following statements are true? **Check the two that apply.**
    a. The pressure relief valve (PRV) is a type of valve used to control or limit the pressure in a vessel or system.
    b. The PRV is designed or set to open at a predetermined set pressure to protect the system.
    c. The fluid (liquid, gas, or liquid–gas mixture) released from the PRV is usually routed through a piping system known as the *blowdown*.
    d. The pressure in a vessel or system typically needs to drop 30% below the predetermined set pressure before the valve resets.

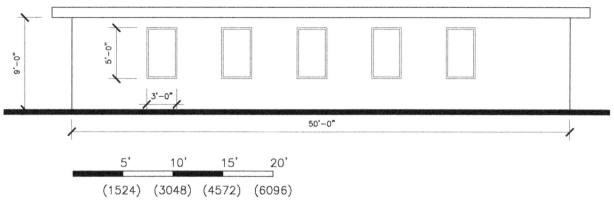

Figure 3.31   South Elevation of a Building

66. Figure 3.31 shows the south elevation of a building. If the U-value for the windows is 0.35, and the R-value for the opaque wall is 21, what is the overall U-value for the entire south wall of the building?
    a. approximately 0.05
    b. approximately 0.08
    c. approximately 0.10
    d. approximately 0.15

67. An architect is working on a theater project in California. There is a conflict between the *ADAAG Manual* and the California Building Code for handicap accessible ramp requirements. The local code has the same requirements as the California Building Code on this issue. What is a proper action for the architect?
    a. Comply with the California Building Code.
    b. Comply with the local code and California Building Code.
    c. Comply with the *ADAAG Manual*
    d. Comply with the most stringent or restrictive code

68. A simple test in construction is to blow smoke through a material. The purpose of this test is to
    a. determine if the material is watertight
    b. determine if the material is airtight
    c. determine if the material is a good sound absorbent
    d. none of the above

69. An architect is designing a federal building with an escalator between first and second floor. Which of the following is true? **Check the two that apply.**
    a. She needs to determine the vertical slope of the escalator.
    b. She needs to use dimensions to locate the work points for the escalator.
    c. She needs to draw and add full dimensions for the escalator so that the contractor can build the escalator accurately.
    d. She needs to coordinate with structural and electrical engineers

70. An architect is trying to locate the main entrance of a CMU building, according to CMU block module, and use either the CMU full block or CMU half block to avoid cutting the CMU blocks. Which of the following dimension fits the CMU block module? **Check the two that apply.**
    a. 18'-4"
    b. 19'-4"
    c. 20'-0"
    d. 20'-4"

71. According to *International Building Code* (IBC), stairways shall typically have a minimum headroom clearance of _____ measured vertically from a line connecting the edge of the nosing.

72. An architect is doing quality control of a set of construction drawings. The Health Department requires all food prep areas have 8' (2438) high FRP. Where are the best places to find the information for the 8' (2438) high FRP? **Check the two that apply.**
    a. Floor plans
    b. Room finish schedules
    c. Interior elevations
    d. Reflected ceiling plans

73. If a set of building plans has to be reviewed by the Health Department, which of the following is likely to be accepted by the Health Department as a floor finish for the Janitor's room? **Check the three that apply.**
    a. Carpet with wood base
    b. Concrete with slim foot base
    c. VCT flooring with cove base
    d. Sheet vinyl flooring with cove base
    e. Ceramic tile floor with cove base
    f. Smooth wood floor with cove base

74. Which of the following are likely to be most cost-effective in North America? **Check the two that apply.**
    a. Concrete over steel deck over steel beams and steel columns framing system
    b. Panelized wood floor over open web steel truss over girder and columns
    c. Panelized wood floor over open web wood truss over girder and columns
    d. Panelized wood floor over purlins over girder and columns

75. Which of the following should not be used in direct contact with wood treated with waterborne preservatives containing copper?
    a. Aluminum
    b. Hot-dipped galvanized steel
    c. Copper
    d. Silicon bronze
    e. Stainless steel

76. Which of the following is the best definition of a horizontal exit?
    a. A horizontal exit is a two-hour separation, separating the building into two compartments.
    b. A horizontal exit is an exit on the same level.
    c. A horizontal exit is an exit enclosed by exit corridor on the same level.
    d. A horizontal exit is always required on every building. It is a basic form of exit.

77. In a building equipped with an automatic sprinkler system throughout, the separation distance of the exit doors or exit access doorways shall not be less than:
    a. One quarter of the length of the maximum overall diagonal dimension of the area served.
    b. One-third of the length of the maximum overall diagonal dimension of the area served.
    c. One-half of the length of the maximum overall diagonal dimension of the area served.
    d. Three quarters of the length of the maximum overall diagonal dimension of the area served.

78. Which of the following are the basic federal laws, which involve accessibility issues? **Check the two that apply.**
    a. Americans with Disabilities Act
    b. Fair Housing Act
    c. Fair Employment & Housing Act
    d. Unruh Civil Rights Act

79. An architect is designing an accessible counter and sinks in a public restroom. Which of the following is correct?
    a. The accessible counter has to be 2'-10" (864) from the finish floor.
    b. The tops of the rims of the sinks, on the accessible counter, have to be 2'-10" (864) from the finish floor.
    c. The accessible counter has to be 2'-10" (864), maximum, from the finish floor.
    d. The tops of the rims of the sinks, on the accessible counter, have to be 2'-10" (864), maximum, from the finish floor.

80. Which of the following statements are not true? **Check the two that apply.**
    a. All accessible counters have to have knee space below the counters.
    b. Accessible counters for workstations have to have knee space below the counters.
    c. Some accessible counters have to have knee space below the counters.
    d. Accessible transaction counters have to have knee space below the counters on the customer side.

81. The term "PVC" is likely to appear in the specifications of which of the following?
    a. Built Up Roof
    b. Single Ply Roof
    c. Tile Roof
    d. Wood Shingle Roof

82. Why is Radon gas not desirable in a building project?
    a. It has bad odor.
    b. It is poisonous.
    c. It is radioactive.
    d. It has too much moisture and is a source of mold problem.

83. An architect is working on a department store. The owner's prototype requires the panic hardware to have a 15-second delay to prevent theft of store merchandise. When the architect submits the plans to the fire department, the plan checker requires the architect to change to a panic hardware without any time delay function. What are the improper actions for the architect? **Check the two that apply.**
    a. Change to a panic hardware without any time delay function per the plan check corrections.
    b. Change to a panic hardware without any time delay function only at the accessible exit doors.
    c. Talk with the plan checker and find out if she can add a sign stating "Keep pushing, the door will open after 15 seconds" as an alternative solution.
    d. Change to a panic hardware without any time delay function per the plan check corrections to get plan check approval, and then tell the owner he can change back to the panic hardware with a 15-second delay after the building is completed.

84. Which of the following toilet partition finishes has the highest initial cost?
    a. Baked enamel
    b. Laminated plastic
    c. Porcelain enamel
    d. Powder shield
    e. Stainless steel
    f. Polly

85. Which of the following normally do not require panic hardware at the required exits? **Check the two that apply.**
    a. Museums
    b. Warehouses with no hazardous materials
    c. Post offices
    d. Restaurants

86. All of the following will affect window selection except:
    a. Building orientation
    b. Location of the window
    c. The low initial cost of low-e glass
    d. The high initial cost of low-e glass

87. The ability of water to flow against gravity through concrete floor cracks is called
    a. Capillary action
    b. Seepage
    c. Saturation
    d. Leakage

88. Right after the punch walk of a project, a HVAC worker falls down through the roof hatch opening and dies. The constructions plans do not show any safety railing around the roof hatch opening. Who is responsible for this accident?
    a. The owner
    b. The Architect
    c. The contractor
    d. The HVAC subcontractor

89. Which of the following is not true about insulation? **Check the two that apply.**
    a. Batt insulation can be attached to the bottom of the roof deck.
    b. Batt insulation can be installed under the concrete slab.
    c. Rigid insulation can be installed under the concrete slab.
    d. Rigid insulation is typically attached to the bottom of the roof deck.

90. Which of the following has the least impact on the design of elevators?
    a. Accessibility
    b. Safety
    c. Number of passengers at peak hour
    d. Building's population

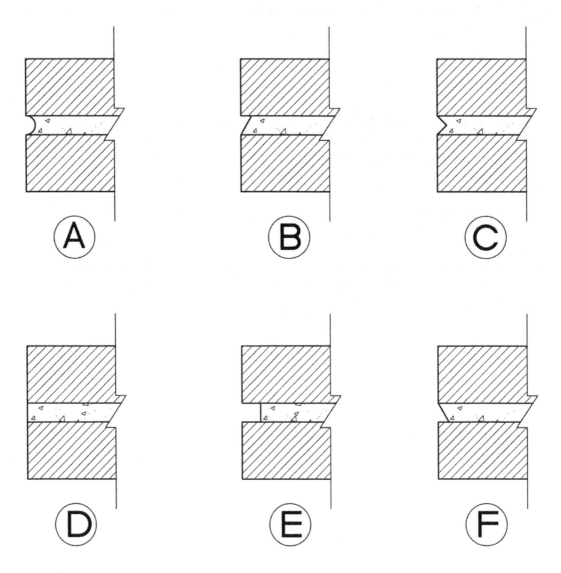

Figure 3.32   Brick joints

91. Which of the previous images shows a Weather Struck brick joint?
    a. A
    b. B
    c. C
    d. D

92. Strips made of white polyethylene are often added to the top and side of glass block partitions. The purpose of the strips is:
    a. To absorb extra moisture from the glass block partition.
    b. To protect the glass block partition in an earthquake.
    c. To provide room for glass block expansion.
    d. None of the above

93. The purpose of a sliptrack on the top of the full-height metal stud walls is:
    a. To protect the metal stud walls from wind forces.
    b. To protect the metal stud walls in an earthquake.
    c. To provide room for metal stud expansion.
    d. All of the above

94. Which of the following are the federal laws that mandate accessibility to certain historic structures? **Check the three that apply.**
    a. Americans with Disabilities Act
    b. Fair Housing Act
    c. Fair Employment & Housing Act
    d. Unruh Civil Rights Act
    e. Architectural Barriers Act
    f. Section 504 of the Rehabilitation Act

95. Which of the following is the best solution to comply with accessibility requirements in a 2-story historic museum?
    a. Install an elevator to provide access to the second floor
    b. Install a lift to provide access to the second floor
    c. Provide a refuge area next to the stair on the second floor
    d. Provide equivalent exhibits on the first floor for handicapped people

96. Which of the following is not recommended when working on the entrance of a historic building?
    a. No documentation on the new work
    b. Stabilizing
    c. Repairing
    d. Replacement

97. The lead based paint for a historic building start to peel, chip, craze, or otherwise comes loose. Which of the following procedure is recommended?
    a. Leave the paint in place and do nothing
    b. Use the same kind of lead-based paint to repaint the damaged area
    c. Remove the lead-paint throughout the building and apply a compatible primer and finish paint
    d. None of the above

98. Which of the following is the most appropriate structural system for a lab building that is sensitive to vibration?
    a. Cast-in-place concrete beam-and-slab system
    b. Heavy timber construction with panelized floor
    c. Lightweight concrete over metal deck over steel joists
    d. 4" gypsum concrete topping slab over wood deck over wood joists

99. Which of the following is appropriate for extinguishing a fire in a cell phone equipment room? **Check the two that apply.**
    a. Dry ice
    b. Water
    c. Dry chemicals
    d. A combination of water, carbon dioxide and dry chemicals

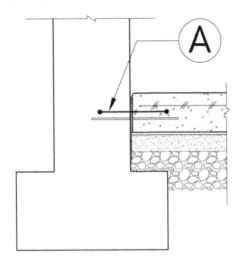

Figure 3.33   The correct term for the preformed synthetic rubber

100. Which of the following is the correct term for the preformed synthetic rubber labeled as letter A in previous image (**Note:** some other elements NOT shown for clarity)?
    a. Waterstop
    b. Moisture barrier
    c. Control joint
    d. Isolation joint

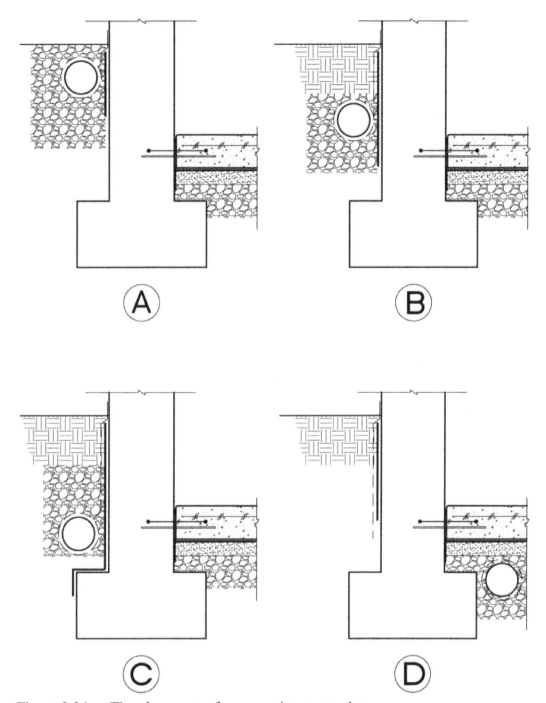

Figure 3.34　The placement of water resistant membrane

101. Which of the previous images shows the correct water resistant membrane of water resistant membrane (shown as heavy solid lines. **Note:** some other elements are NOT shown for clarity)?
    a. A
    b. B
    c. C
    d. D

102. The most important factor for locating a supermarket is
    a. its proximity to potential clients
    b. the availability of utilities
    c. a downtown location
    d. its proximity to high income households

103. A practice or device designed to keep eroded soil on a construction site, so that it does not wash off and cause water pollution to a nearby water body is called
    a. erosion control
    b. sediment control
    c. pollution control
    d. defoliation control

104. A roof overhang on which of the following façades of a building built in the southern hemisphere will provide seasonal adjustment for solar radiation?
    a. North
    b. South
    c. East
    d. West

105. Which of the following are the most important factors in the design of residential units? **Check the two that apply.**
    a. Orientation
    b. The heights and locations of adjacent buildings
    c. Bedrooms facing the dominant wind
    d. Avoiding west facing units

106. Which of the following are correct? **Check the two that apply.**
    a. Detention ponds are also "dry ponds."
    b. Retention ponds are also "dry ponds."
    c. Detention ponds are used to hold stormwater for a short period of time.
    d. Retention ponds are used to hold stormwater for a short period of time.

107. The following are considered environmental impact issues for site analysis EXCEPT **(Check the three that apply.)**
    a. reflections
    b. dominant wind direction
    c. recommended footing design
    d. archeological discoveries
    e. sun and shadow patterns
    f. demography
    g. traffic condition

108. Restrictive covenants are typically controlled by
    a. city
    b. contractor
    c. HOA
    d. EPA

109. Biophilia means
    a. nearsightedness
    b. farsightedness
    c. love at the first sight
    d. human beings' latent desire of being loved
    e. an instinctive bond between human beings and other living systems

110. Looking at the cost of purchasing and operating a building or product, and the relative savings is called
    a. life cycle approach
    b. life cycle assessment (LCA)
    c. life cycle analysis
    d. life cycle costing
    e. life cycle cost and saving analysis
    f. low impact development (LID)

B. Case Study

Questions 111 through 120 refer to the following case study. See figure 3.35 though figure 3.39 for information necessary to answer the questions.

A community library will house special and general collections. Meeting areas for the public will also be available.

1. The site is located to the south of and adjacent to a decorative arts museum. Parking is provided off the site.
2. The major view is to the north.
3. The Lending Desk/Office is to have visual control of the entry to each of the following spaces: the Workroom, the Lobby, and the Children's Reading Room.
4. The main entrance to the building shall be from Fourth Street.
5. Typical ceiling height is 9 ft except the ceiling height for the main reading room is 18 ft.
6. The area of each space shall be within 10% of the required program area.
7. The total corridor area shall not exceed 25% of the total program area.
8. The second floor envelope must be wholly contained within the first floor envelope with the exception that doors to the exterior may be recessed for weather protection.

Program-Spaces			
Tag	Name	Area (sf)	Requirements
ST	Stair	800	2 per floor @ 200 sf per stair
E	Elevator Shaft	200	1 per floor @ 100 sf; Minimum dimension = 7 ft
EE	Elevator Equipment Room	100	
EM	Electrical/Mechanical Room	500	
L	Lobby	350	Main Entrance
LD	Lending Desk/Office	230	
MR	Main Reading Room	2,550	View–exterior window required; 18 ft ceiling; 2 exits; First floor
T	Toilet Rooms	600	2 per floor @ 150 sf each
W	Workroom	400	Exterior windows prohibited; Direct access to Lending Desk/Office
S	Stacks	950	Exterior windows prohibited; Direct access to Main Reading Room
LM	Large Meeting Room	750	Exterior window required; Second floor
SM	Small Meeting Room	350	Exterior window required; Near Large Meeting Room
CR	Children's Reading Room	500	View–exterior window required; Near Main Reading Room
HO	Head Librarian's Office	200	Exterior window required; Direct access to Secretarial Office
AO	Assistant Librarian's Office	150	Exterior window required; Direct access to Secretarial Office
SO	Secretarial Office	300	Exterior window required; Near Large Meeting Room
SC	Special Collections	500	First floor; Exterior window required
B	Break Room	300	Near Large Meeting Room
C	Custodial	200	2 @ 100 sf
TOTAL PROGRAM AREA		**10,200**	

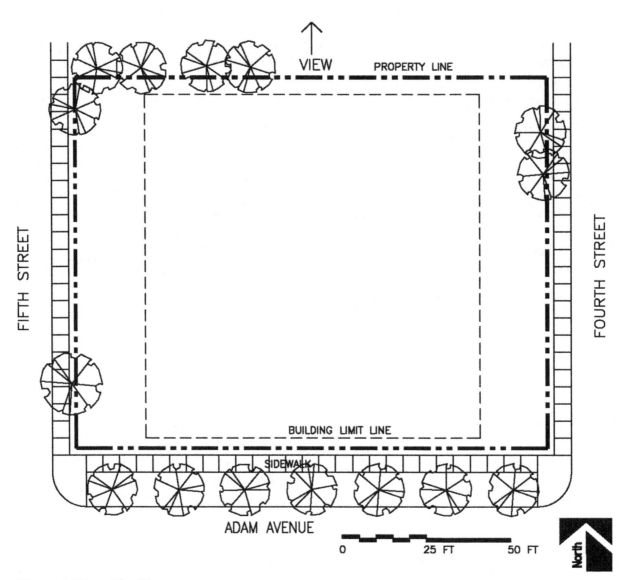

Figure 3.35   Site Plan

91 • Chapter Three

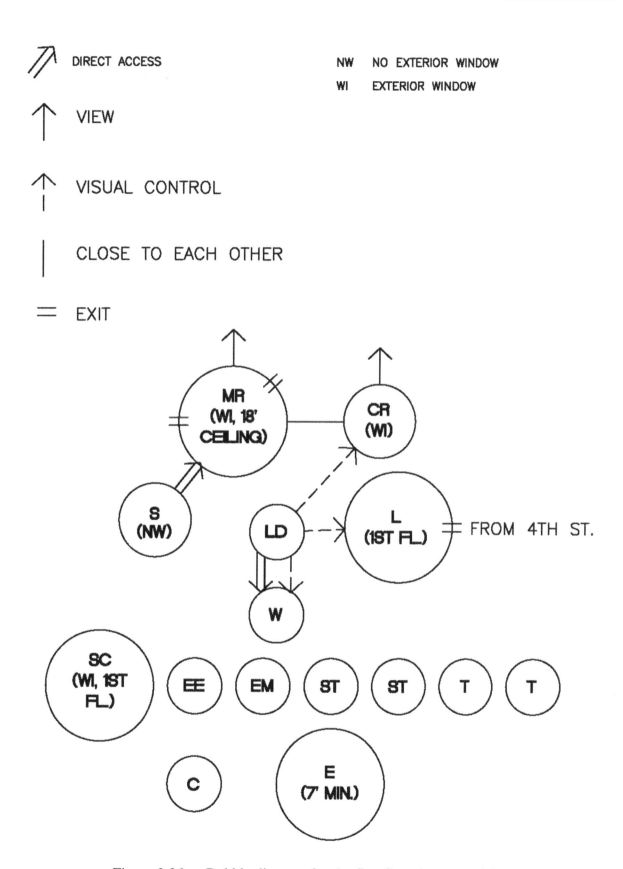

Figure 3.36   Bubble diagram for the first floor (not to scale)

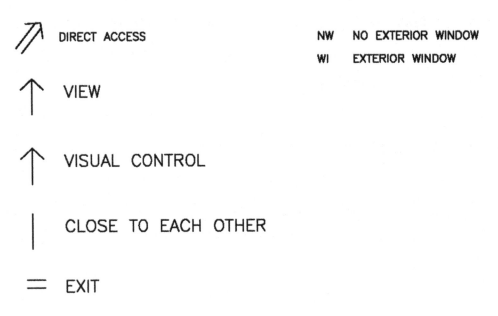

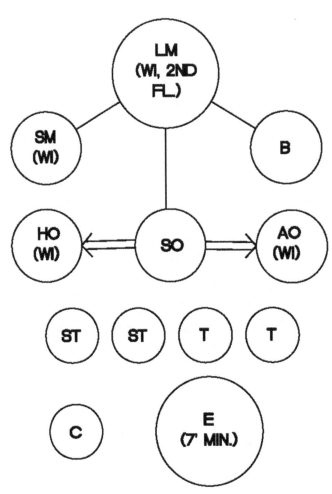

Figure 3.37　Bubble diagram for the second floor (not to scale)

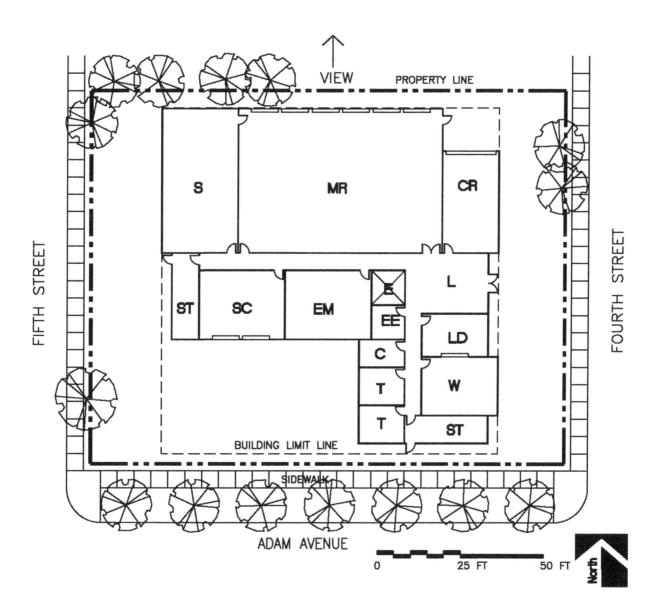

Figure 3.38   Block diagram for the first floor

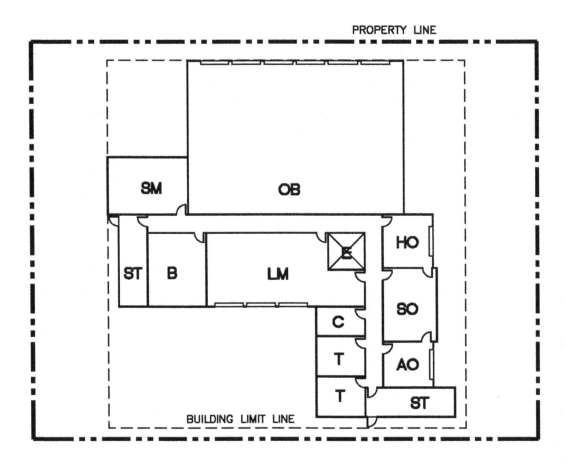

Figure 3.39   Block diagram for the second floor

111. What program requirement is not met in the block diagram?
    a. There is no exterior window for the Stacks.
    b. There is no exterior window for the Small Meeting Room.
    c. There is no view from the Head Librarian's Office.
    d. There is no view from the Assistant Librarian's Office.

112. Which of the following statements regarding the egress system are incorrect? **Check the two that apply.**
    a. Some doors swing in the wrong direction.
    b. Some doors are not necessary.
    c. The exit corridor is not wide enough.
    d. Exit doors should never pass through a stair.

113. Which of the following statements regarding the program requirements are correct? **Check the two that apply.**
    a. Visual control requirements are not completely met.
    b. A room does not have the required direct access.
    c. Toilet Rooms do not have windows per the program.
    d. The Large Meeting Room should have two exits.

114. Drag-and-place the partial sidewalk on the site plan to provide access to a public right of way. The partial sidewalk can be used more than once and at multiple locations.

115. If the client wants to add another 350-SF Meeting Room, where is the best location?
    a. Add it to the north of existing Small Meeting Room, and add a corridor for access.
    b. Add it between the Head Librarian's Office and the Secretarial Office, and expand the second floor to the east.
    c. Add it between the Assistant Librarian's Office and the Secretarial Office, and expand the second floor to the east.
    d. Add it to the north of the Head Librarian's Office, add a corridor for access, and expand second floor to the east.

116. The second floor area is much smaller than the first floor area, which means that the second floor exterior walls do not align with the first floor exterior walls in some rooms. To alleviate this problem, which of the following statements are correct? **Check the two that apply.**
    a. Move Special Collections to the second floor.
    b. Move the Electrical/Mechanical Room to the second floor.
    c. Move the Elevator Equipment Room to the second floor.
    d. Move the Workroom to the second floor.
    e. Expand the second floor to the east, and move the east stair northwards to align the first floor and second floor exterior walls.
    f. There is no way to move any room from the first floor to the second floor.

117. If the roof pitch is 2:12 and there is no parapet, which roofing system should be used?
    a. clay tile roof with one layer of underlayment
    b. concrete tile roof with one layer of underlayment
    c. composition roof with an ice and water membrane throughout the entire roof
    d. single-ply TPO roof
    e. single-ply PVC roof

118. Without extensive re-planning, how should the block diagram be modified to minimize construction cost?
    a. Expand the Head Librarian's Office to align with the first floor exterior wall.
    b. Expand the Secretarial Office to align with the first floor exterior wall.
    c. Expand the Assistant Librarian's Office to align with the first floor exterior wall.
    d. Expand the Head Librarian's Office and the Assistant Librarian's Office to align with the exterior wall of Secretarial Office.

119. What violations of codes and regulations does the block diagram show?
    a. There may be accessibility problems.
    b. There are not enough exits.
    c. The building is too close to the property line and does not have enough setback.
    d. The stair is too narrow and does not have enough exit width.

120. If the client wants to expand the facility later, what should the architect suggest to save on construction costs? **Check the two that apply.**
    a. Expand the second floor to the east to align with the first floor exterior walls.
    b. Expand the second floor to the north to align with the first floor exterior walls.
    c. Expand the building to the south to take full advantage of the site.
    d. Expand the building to the north of the Children's Reading Room.

Chapter Four

ARE Mock Exam Solutions for Project Planning & Design (PPD) Division

A. Mock Exam Answers and Explanations: Multiple-Choice (MC)

Note: If you answer 60% of the questions correctly, you will pass the exam.

1. Answer: a and c
   Per the *International Building Code (IBC)*, the following are correct:
   - *IBC* allows live load reduction in most cases.
   - *IBC* does not allow live load reduction for public assembly occupancy with a live load equal to or less than 100 psf.

   The following are incorrect answers:
   - *IBC* allows live load reduction in a few cases.
   - *IBC* does not allow live load reduction for live load equal to or more than 100 psf.

   See the *International Building Code (IBC)* at following link:
   http://codes.iccsafe.org

2. Answer: a
   Dynamic loads are loads that change rapidly.

   Impact loads are suddenly applied loads.

   Dead loads are vertical loads generated by the weight of the building or permanent equipment.

3. Answer: d. D
   D is the direction of water flow because water always flows from a higher contour line to a lower contour line at a 90-degree angle, or perpendicular to the contour lines.

4. Answer: d. Gravel
   Figure 3.2 shows a typical retaining wall detail with gravels around a perforated pipe. This is to release the hydrostatic pressure behind the retaining all. Neither sand nor soils are proper materials around a perforated pipe because they are too fine and will leak into the perforated pipe.

5. **Answer: c. 113'**
   The spot elevation = 115' - 5' x (10'/25') = 115' - 2' = 113'

6. **Answer: a and d**
   For a *convex* **slope**, the contour lines are closer to each other at the *lower* portion of the slope. A *convex* **slope** has a section profile as shown in Section D, the section of Plan A. Section C is the section of Plan B.

7. **Answer: c**
   If the slope for the lateral sewer line is 1% and the slope for the main sewer line is 1.5%, the invert elevation at point A as shown on figure 3.5 is 95.4'.
   60' x 1% = 0.6
   96.0 - 0.6 = 95.4'

   The following is unnecessary information used as a **distracter** to confuse you:
   The slope for the main sewer line is 1.5%.

8. **Answer: d**
   A cylinder lock is shown on the image. There are four major categories of locks:
   - Mortise lock (fits into **mortise** on the door **edge**)
   - Unit lock (fits into door **cutout**)
   - Integral lock (A **combination** of cylinder lock and mortise lock)
   - Cylinder lock (The most common and cost-effective type, used in many residential buildings. It fits into lock stile of the door and the **drilled holes**)

   See *Building Construction Illustrated* for more information. You need to be able to identify the different kinds of locks when looking at the sketches or images.

9. **Answer: b**
   A pivoting window is shown on the image.
   The major window types by movement include:
   a. Sliding
   b. Pivoting
   c. Casement
   d. Fixed
   e. Awning/Hopper
   f. Double-hung
   g. Jalousie

   You need to be able to identify the window types by reviewing the images or line drawings.

   See *Building Construction Illustrated* for more information:

10. Answer: c
    Letter C in the figure indicates a mullion.

    Letter A in the figure indicates a jamb.

    Letter B in the figure indicates a head.

    Letter D in the figure indicates a sill.

11. Answer: a, b, c and f
    In a building that uses the Insulated Concrete Forms (ICF) system, the forms are left in place permanently for the following reasons:
    - Thermal insulation
    - Acoustic insulation
    - Space to run plumbing pipes and electrical conduits. The form material on either side of the walls can easily accommodate plumbing and electrical installations.
    - Backing for stucco, brick, or other siding on the exterior and gypsum boards on the interior

    Aesthetic effect and construction cost are possible answers, but they are not as good as the other answers.

12. Answer: b
    A **drip edge** is noted as letter A on the image.

    Flashings are used at the intersections of walls, roof, chimney, etc.
    Drip edges are used at the edge of parapets or eaves to drip rainwater.
    Copings are used to cover the top of parapets and may extend to the edge of the parapet.
    Sidings are the finishes for exterior walls.

    *Note: The most important tip for quality control of a project is "**Don't leak and don't fall.**" For the "don't fall" part, you need to pick a good structural engineer, and you need to do a good job to coordinate with him. For the "**Don't leak**" part, you need to make sure ALL your exterior details need to work and keep the water and moisture out of a building. A **drip edge** is an important detail to keep water out of a building.*

13. Answer: a
    A construction joint is noted as letter A on the image.
    - Construction joints also serve as isolation or control joints.
    - Construction joints normally run from the top of slab to the bottom of the slab.
    - Control joints do NOT separate the slab completely, and their depths are normally ¼ of the slab thickness. Control joints can NOT serve as construction joints.
    - Concrete separation and concrete mark are just distractors.

14. Answer: d
    Pattern D on the previous image is stack bond.

    The following are names for all the patterns:
    a. Pattern A: Running bond
    b. Pattern B: Herringbone
    c. Pattern C: Basketweave
    d. Pattern D: Stack bond

15. Answer:
    According to *International Building Code* (IBC), interior adhered masonry veneers shall have a maximum weight of **20** psf (or 97.6 kg/m^2).

    Note: Exterior veneers shall have a maximum weight of 15 psf (or 73.2 kg/m^2).

    These are important numbers to remember. It will help you in selecting the proper masonry veneers in architectural practice. Please note masonry veneers include both **stone** veneers and brick veneers.

    There are two major categories of masonry veneers based on techniques of installation:
    - Adhered masonry veneers (They are easy to install and more cost effective than Anchored masonry veneers)
    - Anchored masonry veneers (They require corrosion-resistant fastenings, metal ties, etc.)

    See Section 1405.10.3 of *International Building Code* (IBC).

    See following links for the FREE IBC code sections citations:
    http://codes.iccsafe.org

    The IBC has a typo:

    "Interior adhered masonry veneers shall have a maximum weight of 20 psf (0.958 kg/ m^2)."

    20 psf (**0.958** kg/ m^2) is a typo, and it should be 20 psf (**97.6** kg/ m^2).

    Here are my calculations:

    1 pound = 0.4536 kilogram

    1 sf = 0.0929 m^2

    20 psf = (20 x 0.4536 kilogram) / 0.0929 m^2 = 97.6 kg/ m^2

16. Answer: b
    Per IBC, the following is correct regarding the panic hardware at the required exits:
    - <u>The actuating portion of the releasing device shall extend at least one-half of the door leaf width.</u>

    See Section 1010.1.10.1 of IBC at the following link:
    http://codes.iccsafe.org

17. Answer:
    Based on the table, the minimum number of water closets required for Women's Restroom for this store is **2**.

    The following is detailed process of the calculations:

Total gross square footage of the building:	32,000 s.f.
Gross square footage for <u>female</u> occupants:	32,000 s.f./2 = 16,000 s.f.
5% of the areas are storage:	5% x 16,000 s.f. = 800 s.f.
The number of the female occupants for storage area: (See Table 3.1)	800 s.f./300 = <u>2.67</u>
The remaining retail spaces:	16,000 s.f.- 800 s.f. =15,200 s.f.
The number of the female occupants for retail spaces:	15,200 s.f./30 = 506.67
Total number of the <u>female</u> occupants:	2.67 + 506.67 = 509.34 or about 509

    Per 2902.1 of IBC, the minimum number of required water closets for Mercantile Occupancy Group is 1 per 500 occupants.

    Therefore, the minimum number of water closets required for Women's Restroom for this store is 509/500 = 1.018 or 2.

    Note: We always round up to the next whole number when calculating the required minimum number of water closets, unless the building official grants an exception. In this case, you can contact the building official and request an exception to allow the minimum number of water closets to be reduced to one. If the building official approves your request, you can then proceed accordingly.

    See Table 1004.1.2 of IBC at the following link:
    http://codes.iccsafe.org

    See Table 2902.1 of IBC at the following link:
    http://codes.iccsafe.org

18. Answer: a
   The image A shows a W shape steel or wide flange.
   The image B shows an S shape steel or American standard I beam.
   The image C shows a T shape steel or structural tee. IT can be cut from a W shape steel or an S shape steel.
   The image D shows a C channel or American standard channel.
   The image E shows an equal leg steel angle (there are also unequal leg steel angles).
   The image F shows a square tube steel (There are also rectangle or round tube steel shapes or pipes).

   There are also combined shapes, such as double angles, a combination section of a C channel and a W shape steel

   See the following books for various basic steel shapes:

   Steel Construction Manual, Latest edition
   American Institute of Steel Construction

   OR

   Handbook of Steel Construction, Latest edition; and *CAN/CSA-S16-01 and CISC Commentary*
   Canadian Institute of Steel Construction

19. Answer:
   Per *International Mechanical Code* (IMC), if the height of a roof access ladder is over **30 feet, an intermediate landing is required.**

   See Section 306.5 of the *International Mechanical Code* (IMC) at the following link: *http://codes.iccsafe.org*

20. Answer:
   The nominal size of a standard brick in the US is Length (L) × Depth (D) × Height (H) = 8" × 4" × 2 5/8" (203 × 102 × 67), and the nominal size of a standard concrete masonry unit (CMU) in the US is H × D x L = 8" × 8" × 16" (203 × 203 × 406). The actual size is usually about 3/8" (10) smaller to allow for mortar joints.

   This means:
   The actual size of a standard brick in the US is 7 5/8" × 3 5/8" × 2 1/4" (193 × 92 × 57), and the actual size of a standard CMU in the US is 7 5/8" × 7 5/8" × 15 5/8" (193 × 193 × 396).

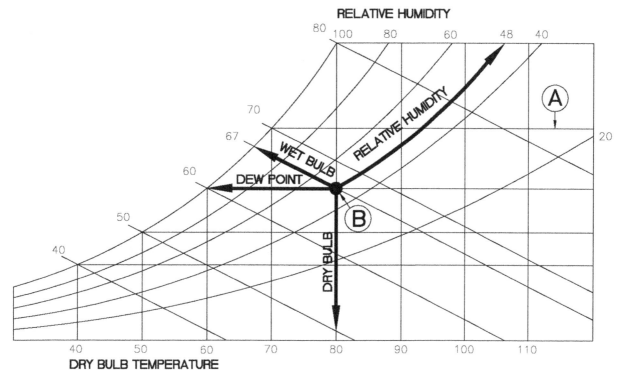

Figure 4.1   A Simple Psychrometric Chart (The temperatures are shown in °F, and the relative humidity is shown in percentages.)

21. Answer: The line labeled as "A" represents <u>a dew point temperature</u>.

Figure 4.1 and 3.13 are simple **psychrometric charts.** Know how to read and use them. Understanding and memorizing what the lines represent is also a good idea.

As long as you know at least two of the properties in the chart, you can determine the others. For example, if you know the **dry** bulb temperature is 80°F and the **wet** bulb temperature is 67°F for a room, you can:
a) Locate the intersection (point "B") of the 80°F **dry** bulb temperature line and the 67°F **wet** bulb temperature line.
b) Follow the curved line from the intersection at point "B" to determine that the **relative humidity** is about 48%. You can use a dehumidifier to reduce the **relative humidity in a room** to 25% to 35% for better human comfort.
c) Determine that the **dew point temperature** is about 60°F by following the horizontal line from point "B". You need to keep the wall **temperature** above 60°F to prevent condensation (figure 4.1).

The example above is to show you the basic methods of how to read and use the chart. There are more detailed and computerized **psychrometric charts** available to assist people to obtaining data that are more accurate. However, the basic principles are the same for all **psychrometric charts**.

22. Answer: 5,000 dollars.
The following is a step-by-step solution:

a) The owner's design criteria require 10,000 BTU per 300 s.f. of gross area (2,930.71 W per 27.87 s.m.).

An office with 1,500 s.f. (139.35 s.m.) of gross area needs:
(1,500 s.f./300 s.f.) x 10,000 BTU = 50,000 BTU (14,653.55 W)

b) In the question, "**ton**" refers to one "**ton of cooling**," a common HVAC unit in North America. This is 12,000 BTU/h or the amount of power needed to melt one **short ton** (2,000 pounds or 907 kilograms or 907 kg) of ice in 24 hours (approximately 3.52 kw).

*Note: Please pay attention to the difference between a **short ton**, a **long ton** and a **metric ton**. A **short ton** is 2,000 pounds or 907 kg; a **long ton** is 2,240 pounds or 1,016 kg; and a **metric ton** is 2,205 lb or 1,000 kg.*

*If a **metric ton** or a **long ton** is used, it is specifically noted.*

50,000 BTU/12,000 BTU = 4.16 ton

c) Since the HVAC system is only available in an integer ton, the smallest HVAC unit we can choose is a 5-ton unit.

The stated market price of a central air-conditioning system is $1,000 per ton, so the minimum cost of a HVAC system for the office that meets the owner's criteria is:

$1,000 x 5 = $5,000 or 5,000 dollars.

Note: This is NOT an easy calculation problem and you may have come up with the right answer for the wrong reasons. We place it at the beginning of the mock exam to test your time management skills. If you spend too much time on this question, you will NOT have enough time to finish the rest of the questions. I suggest you mark ANY calculation problem that requires more than 60 seconds of your time, pick a guess answer, and then move on to other questions. Go back and complete the marked calculation problems AFTER you finish other questions if you have extra time.

This is an important tip for you to pass ANY ARE exam.

23. Answer: b
Figure 3.14 shows a variable flow system. See figure 4.2.
- **A single zone system** delivers conditioned air to various spaces at a constant temperature and low velocity.
- **A variable flow system** uses a damper and thermostat at outlets to control airflow.
- **A double duct system** uses separate ducts to deliver warm and cool air to a terminal mixing box with a damper controlled by a thermostat.

- **A terminal reheat system** provides air at about 55°F (12°C) to terminals with hot water reheat coils.

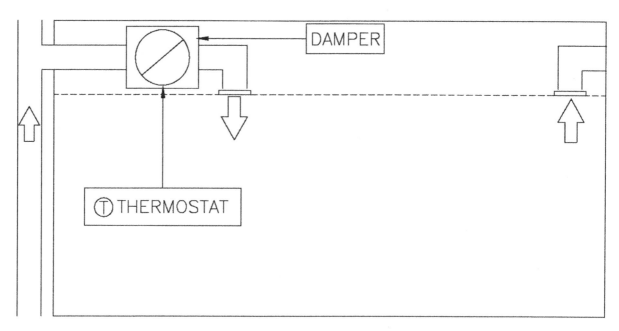

Figure 4.2   Air-Conditioning Distribution System: A Variable Flow System

24. Answer: c
    Capped air chamber is the term for X in figure 3.15. See the explanation to question 12 for further information. X is not an exhaust vent, a manufactured shock absorber, or a vent pipe.

25. Answer: d
    Normally the minimum height of a vent pipe extension is 6" above the roof.
    However, for a roof space accessible by the public or building tenants, the minimum height of a vent pipe extension is 7'-0" above the roof. See section 904.1 of the *International Plumbing Code*.

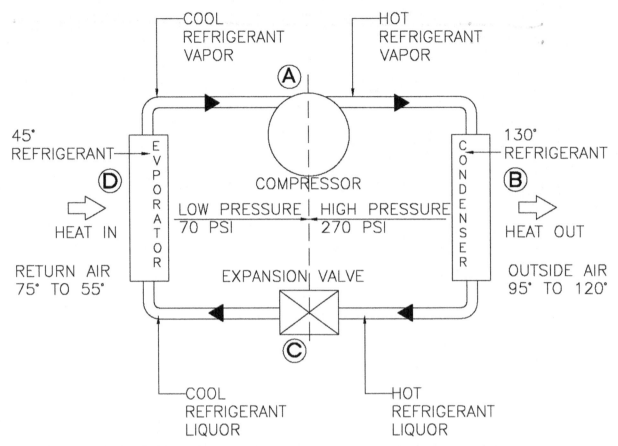

Figure 4.3   Refrigeration Flow Diagram

26. Answer: c
    In the refrigeration flow diagram (figure 3.16 and 4.2), the expansion valve is located at C.

    See the link below for a 1940's U.S. government film on the refrigeration cycle:
    http://www.youtube.com/watch?v=b527al9D_rY

    This video is old but very clearly explains the refrigeration cycle, and the heat transfer processes of conduction, convection, and radiation.
    Similar information and diagrams can be found in *Architectural Graphic Standards (AGS)*.

27. Answer: b and c
    Figure 3.17 indicates the use of two single-pole, double-throw switches and one four-way switch to control lighting from three locations.

28. Answer: b
    The term for Z in figure 3.18 is acoustical lining. It reduces noise buildup inside the machine.

29. Answer: c
Symbol C in figure 3.19 is a floor special purpose outlet.
The correct terms for the other symbols are as follows:
- A: fan hanger receptacle
- B: floor single receptacle outlet
- C: floor special purpose outlet
- D: floor telephone outlet-private

Look through the *Architectural Graphic Standards (AGS)* and become familiar with the symbols used in electrical, mechanical, and plumbing plans.

30. Answer: a
The term for the wiring system in figure 3.20 is single-phase, 3-wire service.

Look through *Mechanical and Electrical Equipment for Buildings* (MEEB), and study all the **images, diagrams, and charts**. Read the related text if you find a diagram or chart confusing.

31. Answer: b
The U-value for the wall is 0.0475 or approximately 0.05.
1) First seek the total resistance of the wall assembly (R_t) by adding together the R-values for each wall component.

Component	R-value
Outside air layer	0.17
3/4" Cement plaster, sand aggregate	0.15
½" Plywood	0.62
Nominal 6" batt fiberglass	19.00
Gypsum board	0.45
Inside air layer	0.68
R_t	21.07

2) The U-value of the wall assembly is then calculated with the following equation:
U = 1/ R_t = 1/21.07 = 0.0475

32. Answer: a
Layout A shows the best orientation of the classroom building to meet the demand of the teachers and students:
- The long side of the building aligns with the east-west axis, and most of the windows face south or north.
- The corridor is placed on the south side. This will not only help block the high angle summer sun, but also allow the low angle winter sun into the classroom. The orientation also yields the greatest potential for outside views with minimum window blind use.
- The windows on the north side can fully utilize natural light with little or no direct sunlight.

33. Answer: c

Figure 3.22 shows a truss. Diagram C shows the correct reactions at A and B.

First of all, support B is on a roller, so the horizontal reaction at support B has to be zero, and the horizontal reaction at support A has to equal 500lb to keep the truss in balance. So, only A and C are possible answers.

Take the moment about support A:
(500 lb x 8 ft) + (V_B x 32 ft) = (500 lb x 32 ft)

Note: V_B is the vertical reaction at support B.

V_B (32 ft) = (500 lb x (32 ft) - (500 lb x 8 ft) = 12,000 lb-ft
V_B = 12,000 lb-ft/32 ft = 375 lb
So, the vertical reaction at support B is 375 lb.

V_A + V_B = 500 lb

Note: V_A is the vertical reaction at support A.

V_A + 375 lb = 500 lb
V_A = 500 lb - 375 lb = 125 lb

So, the vertical reaction at support A is 125 lb.
Therefore, Diagram C is the correct answer.

34. Answer: c

Figure 3.23 shows a building's floor plan. The total wind load in the east-west direction is smaller than in the north-south direction.

The total east and west walls are 60' long, and the north and south walls are 120' long. So, the surface area subject to east-west wind is about half of the surface area subject to north-south wind. Therefore, the total wind load in the east-west direction is smaller than in the north-south direction.

The gust factor is subject to building height and exposure, and not subject to the wind direction. So, answer "d" is incorrect.

35. Answer: a

Figure 3.24 shows beam to column connections. The connection as shown in Diagram I is part of a moment-resisting frame.

To be part of a moment-resisting frame, the beam flange *must* be welded directly to the columns or rigidly attached by plates welded to the columns and bolted to the beam.

The connection, as shown in Diagram II shows a beam seat transferring only the shear from the beam webs to the columns. The top angle is only used to hold the top of the beam in place, and is not a rigid connection to the column.

36. Answer: a and d

Figure 3.25 shows a moment-resisting frame. The following statements are correct:
- The bottom of the column is free to rotate. (There are only two anchor bolts, and they offer little resistance to rotation.)
- The top of the column is fixed against rotation. (This is a moment frame system.)

The top of the column can still translate, or move horizontally.

The following are incorrect answers:
- The top of the column is free to rotate.
- The bottom of the column is fixed against rotation.

37. Answer: d

Diagram D in figure 3.26 shows the distribution of lateral forces used in seismic design. The diagram is normally parabolic. When all stories are the same and the building is short and stiff, the diagram is a reversed triangle, with the maximum lateral force at the top and zero force at the base.

38. Answer: c

The rigid frame shown has moment-resisting connections and fixed bases, so both the upper corners and the two fixed bases are restrained against rotation, as shown in the correct answer, Diagram C.

If the column bases were hinged, the bases would provide no restraint against rotation, and Diagram A would be the correct answer.

If the column bases were hinged *and* the bases could slide, the bases would provide no restraint against rotation or horizontal reactions, making Diagram D the correct answer.

Diagram B is a simple beam supported by columns without rigid joints.

39. Answer: The moment of inertia for the composite section as shown in figure 3.28 is 2049.33 ft-lbf.

The moment of inertia (I) is a measure of the bending stiffness of a structural member's cross-section *shape*. It is the ratio of stress to strain.

For a rectangular section, the moment of inertia about a **horizontal** neutral axis is

$$I = \frac{bd^3}{12}$$

b = the width of the rectangular section
d = the height of the rectangular section

The moment of inertia about a **vertical** axis is

$$I = \frac{bd^3}{3}$$

Without special reference, when we use the term, moment of inertia (I), we typically refer to the moment of inertia about a horizontal neutral axis.

We can transfer the moment of inertia of a composite section's parts to a new centroidal axis using the following equation:

$$I_n = I_x + Ad^2$$

The sum of the transferred moment of inertia is the **composite section's overall moment of inertia**.

Now, let us work on the specificities of this problem.
First, find the centroid of the object using a procedure similar to that given for question 5,
(6" x 10") (13") + (10" x 4") (5") = x (6" x 10"+ 10" x 4")

$x = 9.8$"

We can use a table to simplify our calculations.

Area	I_o (in^4)	A (in^2)	d (in)	Ad^2 (in^4)	$I_o + Ad^2$ (in^4)
A	180	60	3.20	614.40	794.40
B	333.33	40	4.80	921.60	1254.93
The moment of inertia for the composite section = sum of ($I_o + Ad^2$) of Areas A and B					2049.33

$$I_o = \frac{bd_o^3}{12}$$

d_o = width of the part (A or B)
d = distance from centroidal axis to the axes of each part (Area A or B)

111 • Chapter Four

40. Answer: d
    First, convert the loads to load per linear foot of joist:
    Since joists are spaced 4 ft on center, dead load = (30 psf) (4 ft) = 120 plf, live load = (50 psf) (4 ft) = 200 plf, and the total load = 120 plf + 200 plf = 320 plf.

    Second, look at table 3.2 across the 36 ft span row:
    20K10 can support 369 plf of total load, and 193 plf of live load. (Live load is typically shown right below the total).
    22K9 can support 344 plf of total load, and 201 plf of live load.

    Both 20K10 and 22K9 can support the required loads, but 20K10 weighs 12.2 plf, while 22K9 weighs 11.3 plf. So, 22K9 is the most economical section to use.

41. Answer: a
    Figure 3.29 shows a simple truss. Using the method of joints, the force in member AB (the member between joints A and B) is 10,009 lbf.

    Our step-by-step solution is as follows.
 1) Find the **reactions.**
    Since all forces are symmetric, $R_A = R_F = \frac{1}{2}$ (3,000 lbf + 6,000 lbf + 3,000 lbf) = 6,000 lbf.

 2) We presume F_{AB} is a compression force (pointing toward the joint), and F_{AD} is a tension force (pointing away from the joint), and draw joint A as a **free body diagram**.

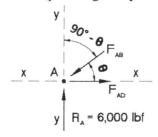

    Figure 4.4   A free body diagram for Joint A

 3) Find the angle between member AB and AD.

$$\tan \theta = \frac{12 \text{ ft}}{16 \text{ ft}}$$

$$\theta = 36.87°$$

    The complementary angle of $\theta = 90° - 36.87° = 53.13°$

 4) The sum of all vertical forces equal zero. Since F_{AB} is acting downward, it is negative:

    $R_A - F_{AB} (\cos 53.13°) = 6,000 \text{ lbf} - F_{AB} = 0$

$$F_{AB} = \frac{6{,}000 \text{ lbf}}{\cos 53.13°} = 10{,}000 \text{ lbf}$$

Note:
Sometimes, you can tilt the x- and y- axes to simplify the calculations. This is an important technique for the method of joints (figure 4.4).

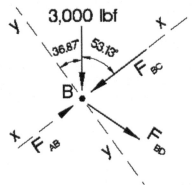

Figure 4.5   A free body diagram for Joint B

42. Answer: a
    In regard to wind load design, the categories B, C, and D are based on **surface roughness**.

    Per IBC, Section 1609.4.2 the surface roughness categories are as follows:
    "**Surface Roughness B.** Urban and suburban areas, wooded areas or other terrain with numerous closely spaced obstructions having the size of single-family dwellings or larger.

    Surface Roughness C. Open terrain with scattered obstructions having heights generally less than 30 feet (9144 mm). This category includes flat open country, grasslands, and all water surfaces in hurricane-prone regions.

    Surface Roughness D. Flat, unobstructed areas and water surfaces outside hurricane-prone regions. This category includes smooth mud flats, salt flats and unbroken ice."

    See IBC Section 1609.4.2 Surface Roughness Categories, for FREE at following link: http://codes.iccsafe.org/I-Codes.html

43. Answer: a
    Diagram A is the correct diagram of wind pressure for a 12-foot high building with a 30-foot square plan.

    There is a formula and several tables you can use to calculate the exact wind loads.

    You do not need to memorize the complicated process of calculating wind loads, but you do need to have a general idea of the directions and distribution of the wind pressures as shown in the diagram and know the basic concepts very well.

For those who are interested in the detailed information, here it is. You just need to look through this, and do not need to memorize.

Per **American Society of Civil Engineers (ASCE)** 7, the following formula can determine wind pressure (P)

P = qGC$_p$ - q$_i$ (GC$_{pi}$)

If a building is symmetric, the previous formula can be simplified as:

P = qGC$_p$

G: This factor takes into account both aerodynamic and atmosphere effect.
G = 0.85 for rigid structures.

C$_p$: This factor takes into account the different effects of the wind on different parts of the building. It can be *negative* when there is negative pressure on the leeward side or roof. You can look it up in *Minimum Design Loads for Buildings* published by ASCE.

Suppose the horizontal distance from the windward edge is *d*, and building height is *h*.

When *d = 0 to h*, **C$_p$** = -0.9

When *d = h to 2h*, **C$_p$** = -0.5

When *d ≥ h to 2h*, **C$_p$** = -0.3

When **C$_p$** is negative, it means the wind pressure is acting away from the building surface.

Since **P = qGC$_p$**, and **q** and **G** are the same for the entire flat roof, **C$_p$** determines the **P** on the flat roof. When the P is shown graphically, it looks like diagram A.

C$_{pi}$ is a factor for evaluating internal pressures.

C_{pi} = ±0.18 for a *completely* enclosed building.

C_{pi} = ±0.55 for a *partially* enclosed building.

If a building is symmetric, the internal pressures on opposing walls will cancel out.

q is the wind stagnation factor.

q = 0.00256K$_z$K$_{zt}$K$_d$v^{2}I

K$_z$ takes into account the combined effects of exposure, height, and wind gust. You can look it up in ASCE 7, table 6-3.

K_{zt} takes into account escarpments and hills close to the building.
$K_{zt} = 1$ for level ground.

K_d takes into account the type of structure being studied.
$K_d = 0.85$ for buildings

I is the importance factor.
You can look this up in *Minimum Design Loads for Buildings* published by ASCE.

Note:
A building is to be classified as one of the following:

open building
- $A_o \geq 0.8 A_g$ for each wall. A_o is the total area of the openings in the wall, and A_g is the total gross area of the wall.

partially enclosed building
- $A_o > 1.10 A_{oi}$, and
- $A_o >$ minimum 4 sf, or $> 0.01 A_g$, whichever is smaller
- $A_{oi}/A_{gi} \leq 0.20$

  A_o and A_g have been defined earlier.

  A_{oi} is the total area of the openings in the building envelope (walls and roof), NOT including A_o.

  A_{gi} is the total gross area of the building envelope (walls and roof), NOT including A_g.

enclosed building
- *This is a building that is neither open nor partially enclosed.*

44. Answer: b and d
    The following statements regarding the Seismic Design Category (SDC) of a structure are true:
    - Buildings in SDC C need to meet more restrictive earthquake requirements than buildings in SDC A.
    - SDC includes categories A through E.

    Occupancy group I are buildings with a *low hazard* to human life if they fail, such as agricultural facilities, minor storage facilities, and temporary facilities.

    Occupancy group II are other buildings *not* included in occupancy groups I, III, and IV.

Occupancy group III are buildings with *a substantial hazard* to human life if they fail, such as public assemblies with an occupant load over 300; elementary or secondary schools, or daycare facilities with an occupant load over 250; colleges or adult education facilities with an occupant load over 500; jail and detention facilities; power-generating, potable water-treatment facilities, waste water-treatment facilities, or other public facilities not included in occupancy group IV; buildings with sufficient quantities of explosives or toxic substances and not included in occupancy group IV; or health care facilities with 50 or more resident patients, but without surgery or emergency treatment facilities.

Occupancy group IV are buildings designated as *essential facilities*, such as hospitals or health care facilities with surgery or emergency treatment facilities; police, fire, and rescue stations or emergency vehicle garages; designated earthquake, hurricane, or other emergency shelters; designated communication, emergency preparedness, and operation centers for emergency responses; power-generating or other public facilities required as back up facilities for occupancy group IV structures; buildings with highly toxic substances exceeding the maximum allowable quantities of Table 307.1.(2); air traffic control centers, aviation control towers, or emergency aircraft hangers; structures with significant national defense functions; or water-treatment facilities required to maintain water pressure for fire suppression.

Seismic Design Category A are buildings of **all** occupancy groups in areas expecting minor ground shaking. (Good soil)
Seismic Design Category B are buildings of occupancy groups I, II, and III in areas expecting moderate ground shaking. (Stratified soil with good and poor soil)
Seismic Design Category C are buildings of occupancy groups IV in areas expecting moderate ground shaking, and buildings of occupancy categories I, II, and, III in areas expecting SEVERE ground shaking.
Seismic Design Category D are buildings of **all** occupancy groups in areas expecting severe and destructive ground shaking, but NOT located close to a major fault. (Sites with poor soil are a good example.)
Seismic Design Category E are buildings of occupancy groups I, II, and III in areas near major active faults. (Soil or rock is of no consequence.)
Seismic Design Category F are buildings of Occupancy Groups IV in areas near major active faults. (Soil or rock is of no consequence.)

The following are incorrect answers:
- Buildings in SDC A need to meet more restrictive earthquake requirements than Buildings in SDC C.
- SDC includes categories A through D.

45. Answer: a and c
Rainwater and water from laundry tubs and clothes washers is **graywater**.

Water from dishwashers, kitchen sinks, and toilets is **blackwater**. See explanation to answer 15 for further information.

46. Answer: c
Underground water lines should be placed below the frost line to avoid freezing.

47. Answer: b
The solar altitude is largest in the Northern Hemisphere on the day of the summer solstice and smallest on the day of winter solstice. Summer solstice is the day of the year with the longest period of daylight, while winter solstice is the day of the year with the shortest period of daylight.

These days are exactly the opposite in the Southern Hemisphere.

On the equinox (vernal and autumnal), night and day are roughly the same lengths.

48. Answer: a
She should submit a Conditional Use Permit application to the city.
- **Non-conforming Use**: the use that does not conform to the current zoning ordinance, but allowed to stay or grandfathered in.
- **Incentive Zoning**: Use the permission to build a larger or taller building as an incentive for a private developer to provide public amenities such as a park.
- **Ordinance Variance**: This is just a **distracter**.

49. Answer: a and d

Albedo is a word derived from Latin *albedo* "whiteness" (or reflected sunlight). It means the ratio of reflected sunlight. The albedo of a perfectly white surface is 1, and the albedo of a perfectly black surface is 0.

The following statements are correct:
- In a cold climate, an architect should use materials with low albedo and low conductivity for ground surfaces. These surfaces will absorb more sunlight and retain the heat as close to the surfaces as possible.
- In a tropical climate, an architect should use materials with high albedo and high conductivity for ground surfaces. These surfaces will absorb less sunlight and quickly absorb and store the heat, and as quickly released when the temperature drops. This will help to produce a mild and stable microclimate.

The following statements are incorrect:
- In a cold climate, an architect should use materials low with albedo and high conductivity for ground surfaces.
- In a tropical climate, an architect should use materials with low albedo and low conductivity for ground surfaces.

50. Answer: Per EPA, using a watershed-based approach to wetland protection ensures that the whole system, including land, air, and water resources, is protected.

    See Wetland information at the EPA website:
    https://www.epa.gov/wetlands

51. Answer: a, b, d, and f
    Per EPA, the following are general categories of wetlands found in the US:
    - Bogs
    - Fens
    - Marshes
    - Swamps

    Lakes and Reservoirs are just distracters.

    According to EPA:
    "**Marshes** are wetlands dominated by soft-stemmed vegetation, while **swamps** have mostly woody plants. **Bogs** are freshwater wetlands, often formed in old glacial lakes, characterized by spongy peat deposits, evergreen trees and shrubs, and a floor covered by a thick carpet of sphagnum moss. **Fens** are freshwater peat-forming wetlands covered mostly by grasses, sedges, reeds, and wildflowers."

    See Wetland information at the EPA website:
    https://www.epa.gov/wetlands

52. Answer: c
    Per EPA, habitat degradation is the leading cause of species extinction.

    Human activities, pollution and excessive hunting all play a role in species extinction, but they are not the leading cause.

53. Answer: c and d
    An architect is working on a residential project. The following are correct statements:
    - The architect should contact the Planning Department to obtain site coverage ratio.
    - The architect should contact the Planning Department to obtain zoning information for the site.

    The following are incorrect answers:
    - The architect should review the building codes for FAR information (She should contact the Planning Department to obtain FAR information).
    - The architect should contact the Planning Department to obtain information on the existing easement and set back requirements for the front yard, the side yard and the back yard (She should contact the owner for easement information. The owner should have received a title report from the title company, and the title report should have the easement information).

54. Answer: c and e
    Pay attention to the word "not."

    Putting acoustical lining outside of the HVAC ducts or using stainless steel ducts are <u>not</u> effective ways to control noise between two adjacent apartment units, and are therefore the correct answers.

    However, flexible boots, resilient hangers and acoustical lining inside of the HVAC ducts are effective ways to control noise between two adjacent apartment units, and are therefore the incorrect answers.

55. Answer: a and c
    The following statements are true:
    - Duct silencers and baffles are normally placed inside the HVAC ducts.
    - Duct silencers and baffles are useful to reduce fan noise but cause considerable pressure drop.

56. Answer: b
    The tenant needs fourteen (14) recessed light fixtures for this office.

    Based on the NCARB Building Systems division list of references and formulas sheet (BSreferences.pdf, available in the real exam) and our discussions on page 24, we can do the following calculations:

    Lumens per fixture= (lumens per lamp) x (number of lamps per fixture) = 2,800 x 4 =11,200

    Number of 2'x4' recessed light fixtures = Number of luminaires = (footcandles) x (floor area) / (lumens per fixture) x (CU) x (LLF) = 50 x 1,200 /11,200 x 0.60 x 0.65 =13.73 or about 14.

57. Answer: a
    The tenant needs twenty-six (26) 1'x4' light fixtures for this office.

    Lumens per fixture= (lumens per lamp) x (number of lamps per fixture) = 3,000 x 2 = 6,000

    Number of 1'x4' light fixtures = Number of luminaires = (footcandles) x (floor area) / (lumens per fixture) x (CU) x (LLF) = 50 x 1,200 /6,000 x 0.60 x 0.65 = 25.64 or about 26.

58. Answer: c
    The tenant utility cost for this office is $25 per month.
    Footcandles [lux] = lumens / area in s.f. (s.m.)
    Lumens = footcandles [lux] x [area in s.f. (s.m.)] = 50 x 1,200 = 60,000
    Utilities cost per year = (60,000/1,000) x 5 = $300
    Utilities cost per month = Utilities cost per year/12 = 300/12 = $25

119 • Chapter Four

59. Answer: b
Since the tenant needs fourteen (14) recessed light fixtures for this office, the tenant cost for installing all the lights in this office = 14 x $300 = $4,200.

60. Answer: b, c, and e
The architect can find electrical service information on the:
- single-line diagram
- panel schedules
- power plans

Electrical lighting plans and low voltage plans do not contain electrical service information on them.

As an architect, you need to know enough about your consultants' work to be able to coordinate with them.

61. Answer: a, b, and c
The architect should forward the mechanical roof plan and HVAC equipment schedules received during design development to the following project team members for coordination:
- The plumbing engineer provides condensation lines for the rooftop HVAC units.
- The electrical engineer supplies power for the rooftop HVAC units.
- The structural engineer makes sure the structure can support the rooftop HVAC units.

The architect should NOT forward the information to the following project team members:
- The contractor does not get involved until the project is out to bid for a design-bid-build project.
- The civil engineer does not need rooftop HVAC unit information.
- The fire protection engineer does not need to coordinate the rooftop HVAC units for typical buildings.

62. Answer: c
The most cost-effective way to achieve this goal is to place the additional HVAC equipment on a concrete pad at grade outside of the building and install new ducts to this equipment.

Adding additional columns and beams to support the additional weight of the new HVAC equipment is NOT cost-effective.

Having an X-ray taken of the building structure to find out if it can support the additional weight of the new HVAC equipment is very expensive.

Placing the new HVAC equipment right on top of the existing structural columns may not provide adequate structural support, and the columns may conflict with the HVAC supply or return ducts.

63. Answer: a
A **fan coil unit (FCU)** is a simple device consisting of a heating or cooling coil and fan. A FCU is part of an HVAC system, and typically NOT connected to ductwork. Either a thermostat or a manual on/off switch controls the device.

64. Answer: a and d
The following statements are true:
- A dry pipe fire sprinkler system is one in which the pipes are filled with pressurized air, rather than water.
- A dry pipe fire sprinkler system will not freeze in unheated spaces.

The following statements are false:
- A dry pipe fire sprinkler system is one in which the pipes are filled with pressurized Halon, rather than water.
- A dry pipe fire sprinkler system is lighter and less expensive to install than a wet-pipe sprinkler system
- A dry pipe fire sprinkler system has fewer valves and fittings to maintain.

65. Answer: a and b
The following statements are true:
- The pressure relief valve (PRV) is a type of valve used to control or limit the pressure in a vessel or system.
- The PRV is designed to open at a predetermined set pressure to protect the system.

The following statements are false:
- The fluid (liquid, gas, or liquid–gas mixture) released from the PRV is usually routed through a piping system known as the *blowdown*. (The *blowdown* is a percentage of set pressure and refers to how much the pressure needs to drop before the valve reseats.)
- The pressure in a vessel or system typically needs to drop 30% below the predetermined set pressure before the valve resets. (The pressure in a vessel or system typically needs to drop 2% to 20% below the predetermined set pressure before the valve reseats.)

66. Answer: c
The overall U-value for the south wall of the building is approximately 0.10.
$U_w = 0.35$
$R_{op} = 21$
$U_{op} = 1/21 = 0.048$

$A_w = (3'-0" \times 5'-0") \times 5$ windows $= 75$ s.f.
$A_o = 9'-0" \times 50'-0" = 450$ s.f.
$A_{op} = A_o - A_w = 450$ s.f. $- 75$ s.f. $= 375$ s.f.

Per the reference sheet provided by NCARB:
$U_o = [(U_w \times A_w) + (U_{op} \times A_{op})]/A_o = [(0.35 \times 75) + (0.048 \times 375)]/450 = 0.098$ or approximately 0.10

67. Answer: d
The architect should comply with the most stringent or restrictive code when there is a conflict between the federal, state (provincial), or local code.

68. Answer: c
A simple test in construction is to blow smoke through a material. The purpose of this test is to determine if the material is a good sound absorbent. If the material is porous, thick, fibrous, and allows smoke to pass freely, it should be a good sound absorbent.

69. Answer: b and d
The following are true:
- She needs to use dimensions to locate the work points for the escalator. (An architect needs to refer to escalator manufacturer's brochure, specify the model of the escalator, locate the two work points for the escalator, and make sure there is adequate clearance and headroom below the escalator. The contractor can show the remaining dimensions as part of the shop drawings or submittals.)
- She needs to coordinate with structural and electrical engineers.

The following are not true:
- She needs to determine the vertical slope of the escalator. (The vertical slope of an escalator is always set at 30 degree. An architect does NOT determine it)
- She needs to draw and add full dimensions for the escalator so that the contractor can build the escalator accurately. (She needs to use dimensions to locate the work points for the escalator, but she does NOT need to draw and add full dimensions for the escalator. The contractor does the full dimensions as part of the shops drawings or submittals.)

70. Answer: b and c
Both 19'-4" and 20'-0" fit the CMU block module dimension. The nominal size of a standard concrete masonry unit (CMU) in the US is 8" × 8" × 16" (203 × 203 × 406). Most CMU manufacturers also produce half blocks at the standard size of 8" × 8" × 8" (203 × 203 × 203).

(18'-4") / 8" = 27.5   (This dimensions is not according to CMU block module, and CMU blocks have to be cut.)

(19'-4") / 8" = 29   (This dimensions is according to CMU block module, and CMU blocks do NOT have to be cut.)

(20'-0") / 8" = 30   (This dimensions is according to CMU block module, and CMU blocks do NOT have to be cut.)

(20'-4") / 8" = 30.5   (This dimensions is not according to CMU block module, and CMU blocks have to be cut.)

Note: There is a simple rule of thumb to determine if a dimension meets the CMU block module:

*If the dimension has an **odd** number on the feet unit, and you **end up with 4"**, it meets the CMU block module. For example, 1'-4", 3'-4", 5'-4", 7'-4", 9'-4", and 11'-4" all meet the CMU block module.*

*If the dimension has an **even** number on the feet unit, and you **end up with 0" or 8"**, it meets the CMU block module. For example, 2'-0", 2'-8", 4'-0", 4'-8", 6'-0", and 6''-8" all meet the CMU block module.*

71. Answer:
    According to *International Building Code* (IBC), stairways shall have a minimum headroom clearance of 80" (2032) measured vertically from a line connecting the edge of the nosing.

    See Section 1003.2 and Section 1011.3 of *International Building Code* (IBC) at the following link:
    http://codes.iccsafe.org

72. Answer: b and c
    The best places to find the information for the 8' high FRP are room finish schedules and interior elevations because the 8' high FRP are interior room finish for the food prep areas.

73. Answer: b, d and e
    If a set of building plans has to be reviewed by the Health Department, the following are **likely** to be acceptable by the Health Department as floor finish for the Janitor's room because they can be cleaned easily:
    - Concrete with slim foot base
    - Sheet vinyl flooring with cove base
    - Ceramic tile floor with cove base

    The following are **unlikely** to be acceptable by the Health Department as floor finish for the Janitor's room because they cannot be cleaned easily, or are NOT tolerant of water and moisture:
    - Carpet with wood base
    - VCT flooring with cove base (VCT stands for vinyl composition tile)
    - Smooth wood floor with cove base

74. Answer: b and c
    The following is likely to be most cost-effective in North America:
    - Panelized wood floor over open web steel truss over girder and columns
    - Panelized wood floor over open web wood truss over girder and columns

123 • Chapter Four

The following is unlikely to be most cost-effective in North America:
- Concrete over steel deck over steel beams and steel columns framing system
- Panelized wood floor over purlins over girder and columns

In North America, concrete over steel deck is typically more expensive than panelized wood floor. Panelized wood floor over purlins is typically more expensive than panelized wood floor over open web wood truss or panelized wood floor over open web steel truss.

75. Answer: a
Waterborne preservatives can increase the risk of corrosion when metals contact treated wood used in wet locations.

Aluminum should not be used in direct contact with wood treated with waterborne preservatives containing copper. On the other hand, hot-dipped galvanized steel, copper, silicon bronze, or stainless steel can be used in direct contact with wood treated with waterborne preservatives containing copper

76. Answer: a
A **horizontal exit** is basically a **two-hour separation**, separating the building into two compartments, A and B. This two-hour separation needs to extend from exterior wall to exterior wall and separate the two parts of building completely. **When people from compartment A enter compartment B, they "exited" from compartment A**, and vice versa. Horizontal exits are extremely valuable for assembly occupancies like casino or convention spaces or a large restaurant on the top story of a high-rise building because they can avoid the use of huge exit stairs, and save spaces.

A horizontal exit is an exit on the same level, but its most important feature is the two-hour separation.

A horizontal exit is NOT an exit enclosed by exit corridor on the same level.

A horizontal exit is NOT always required on every building. It is NOT a basic form of exit.

77. Answer: b
In a building equipped with an **automatic sprinkler** system throughout, the separation distance of the exit doors or exit access doorways shall not be less than **one-third** of the length of the maximum overall diagonal dimension of the area served.

If the building is **NOT** equipped throughout with an automatic sprinkler system, the separation distance of the exit doors or exit access doorways shall not be less than **one-half** of the length of the maximum overall diagonal dimension of the area served.

See Section 101.2, exception #2 of *International Building Code* (IBC) at the following link:
http://codes.iccsafe.org

78. Answer: a and b
The following are the basic federal laws involve accessibility issues:
- Americans with Disabilities Act
- Fair Housing Act

The following are the basic California laws involve accessibility issues:
- Fair Employment & Housing Act
- Unruh Civil Rights Act
- Disabled Persons Act

79. Answer: d
An architect is designing an accessible counter and the sinks in a public restroom. The tops of the rims of the sinks on the accessible counter have to be 2'-10" (864) maximum from the finish floor. It can be a little lower, and does not have to be 2'-10" (864) maximum from the finish floor.

See more information at the following link:
http://www.access-board.gov/

80. Answer: a and d
Pay attention to the word "not." We are looking for the statements that are not true.

The following statements are true, and therefore the incorrect answers:
- Accessible counters for workstations have to have knee space below the counters.
- Some accessible counters have to have knee space below the counters.

The following statements are untrue, and therefore the correct answers:
- All accessible counters have to have knee space below the counters. (NOT all accessible counters have to have knee space below the counters. For example, the accessible transaction counters in a store do NOT have to have knee space below the counters on the customer side.)
- Accessible transaction counters have to have knee space below the counters on the customer side. (Accessible counters for transaction stations do NOT have to have knee space below the counters on the customer side.)

81. Answer: b
The term "PVC" is likely to appear in the specifications of Single Ply Roof.

There are two major categories of membrane roofing systems:
- Bituminous Systems (the traditional systems)
- Single Ply Roof Systems (the newer systems)

They can be further divided as following:

Built Up Roof is one subcategory of Bituminous Systems. Bituminous Systems includes three basic subcategories:

- Bituminous Systems - BUR (Built Up)
- Bituminous Systems - APP Modified Bitumen
- Bituminous Systems - SBS Modified Bitumen

Single Ply Roofing Systems includes several basic subcategories:
- Single Ply - TPO (an ethylene propylene rubber)
- Single Ply - PVC (a thermoplastic material)
- Single Ply – EPDM (an elastomeric material)
- Single Ply – CSPE (a synthetic rubber)
- Single Ply – Neoprene (a synthetic rubber)
- Single Ply – Polymer-modified Bitumen (a composite material)

You need to be familiar with these terms.

Tile Roofing and Wood Shingle Roofing are distractors.

82. Answer: c
Radon gas is not desirable in a building project because it is radioactive and is considered a health hazard.

Radon is atomic number 86, a chemical element with symbol Rn. It is an odorless, colorless, radioactive, tasteless noble gas, occurring naturally as the decay product of uranium (atomic number 92). It is considered a health hazard due to its radioactivity.

The following statements are not true:
- It has bad odor.
- It is poisonous.
- It has too much moisture and is a source of mold problem.

83. Answer: b and d
Please note we are looking for **improper** actions for the architect.

The following statements are improper actions for the architect and therefore the correct answer:
- Change to a panic hardware without any time delay function only at the accessible exit doors. (This does not work because the plan check corrections apply to ALL exit doors and the related panic hardware.)
- Change to a panic hardware without any time delay function per the plan check corrections to get plan check approval, and then tell the owner he can change back to the panic hardware with a 15-second delay after the building is completed. (This is unethical and against the NCARB **Rules of Conduct**.)

Rules of Conduct is Available as a FREE PDF file at (Skimming through it should be adequate):
http://www.ncarb.org/

The following statements are proper actions for the architect and therefore the incorrect answer:
- Change to a panic hardware without any time delay function per the plan check corrections.
- Talk with the plan checker and find out if she can add a sign stating "Keep pushing, the door will open after 15 seconds" as an alternative solution.

84. Answer: e
Stainless steel toilet partition finishes has the highest initial cost.
Baked enamel has the lowest initial cost. Powder shield is powder coated baked enamel. Polly is solid plastic HDPE. Porcelain enamel has a low initial cost also.

85. Answer: b and c
Please pay attention to the word "not" in the original question.
The following normally do not require panic hardware at the required exits, and therefore are the correct answers:
- Warehouses with no hazardous materials
- Post offices

The following normally require panic hardware at the required exits, and therefore the incorrect answers:
- Museums
- Restaurants

Per Section 1010.1.10 of IBC, each door in a means of egress from a Group A or E occupancy having an occupant load of 50 or more and any Group H occupancy shall not be provided with a latch or lock unless it is panic hardware or fire exit hardware.

Per Section 303 of IBC, museums and restaurants belong to Assembly Group A, and require panic hardware at the required exits.

Per Section 304 of IBC, post offices belong to Business Group B, and do not require panic hardware at the required exits.

Per Section 311 of IBC, warehouses with no hazardous materials belong to Storage Group S, and do not require panic hardware at the required exits.

See following link for the FREE IBC code sections citations:
http://codes.iccsafe.org

86. Answer: c
Pay attention to the word "except."

The low-e glass has a visible light transmission value, and a low heat transfer coefficient. The low-e glass has a high initial cost, but is energy efficient, and can save the owner money over a long-term.

All of the following will affect window selection, and are therefore the incorrect answers:
- Building orientation
- Location of the window
- The high initial cost of low-e glass

87. Answer: a
The ability of water to flow against gravity through concrete floor cracks is called a **capillary action**.

Seepage is the slow escape of a liquid or gas through porous material or small holes.

Saturation is the state or process that occurs when no more of something can be absorbed, combined with, or added.

Leakage is the accidental admission or escape of a fluid or gas through a hole or crack.

88. Answer: b
The architect is responsible for this accident because the constructions plans do not show any safety railing around the roof hatch opening.

Note: Many architects miss the roof hatch safety railing but BOTH OSHA Standards 29 CFR 1910.23 and 29 CFR 1910.27 require it.

Make sure you show a standard roof hatch safety railing. It can be a simple note calling out the manufacturer and model number and a simple graphic on your roof hatch detail.

I know at least three manufacturers who make them. See the links below:
http://www.simplifiedbuilding.com/keehatch.php
http://www.freepatentsonline.com/6681528.html
http://www.4specs.com/s/07/07-7230.html

89. Answer: b and d
Please note we are looking for statements that are NOT true.

The following statements are not true and therefore the correct answers:
- Batt insulation can be installed under the concrete slab. (Batt insulation cannot be installed under the concrete slab. It can be installed under the roof deck, or right above the suspended ceiling.)
- Rigid insulation is typically attached to the bottom of the roof deck. (Rigid insulation is typically installed above the roof deck.)

The following statements are true, and therefore the incorrect answers:
- Batt insulation can attached to the bottom of the roof deck.
- Rigid insulation can be installed under the concrete slab.

90. Answer: d
    Building's population has the least impact on the design of elevators.

    Accessibility and safety are mandatory requirements for the design of elevators

    Number of passengers at peak hour will help the designer to determine the number and size of the elevators needed.

    Building's population has some influences on the design of elevators, but is has the least impact on the design of elevators.

91. Answer: b
    The image B shows a Weather Struck brick joint.

    The image A shows a Concave brick joint.
    The image C shows a Vee brick joint.
    The image D shows a Flush brick joint.
    The image E shows a Raked brick joint.
    The image F shows a Trowel Struck brick joint.

92. Answer: c
    Strips made of white polyethylene are often added to the top and side of glass block partition. The purpose of the strips is to provide room for glass block expansion. The strips are called **expansion strips.**

    The following are incorrect answers:
    - To absorb extra moisture from the glass block partition.
    - To protect the glass block partition in an earthquake. (There is some truth to this, but it is NOT the best answer.)
    - None of the above

93. Answer: d
    The correct answer is "all of the above."

    The purpose of sliptrack or slotted deflection track on the top of the full-height metal stud walls is:
    - To protect the metal stud walls from wind forces.
    - To protect the metal stud walls in an earthquake.
    - To provide room for metal stud expansion and contraction.

94. Answer: a, e and f
    The following are the federal laws that mandate accessibility to certain historic structures:
    - Americans with Disabilities Act
    - Architectural Barriers Act
    - Section 504 of the Rehabilitation Act

129 • Chapter Four

<u>Fair Housing Act is the basic federal laws involve housing accessibility issues:</u>

The following are the basic California laws involve accessibility issues:
- Fair Employment & Housing Act
- Unruh Civil Rights Act

See page 14 of the PDF file for *The Secretary of the Interior's Standards for the Treatment of Historic Properties with Guidelines for Preserving, Rehabilitating Restoring & Reconstructing Historic Buildings.* You can simply Google with its name and find the FREE pdf file.

95. Answer: d
Providing equivalent exhibits on the first floor for handicap people is the best solution to comply with accessibility requirements in a 2-story historic museum. This solution will NOT alter the historic museum.

Providing a refuge area next to the stair on the second floor will only provide refuge area for emergency, and does not provide accessibility for handicap people.

All the other solutions listed below will alter the historic museum, and therefore are not the correct answers:
- Install an elevator to provide access to the second floor
- Install a lift to provide access to the second floor

96. Answer: a
Not having documentation on the new work is not recommended when working on the entrance of a historic building.

The following are all acceptable:
<u>Stabilizing</u> <u>deteriorated or damaged entrances and porches as a preliminary measure.</u>

<u>Repairing</u> <u>entrances and porches by reinforcing the historic materials using recognized preservation methods.</u>

<u>Replacement:</u> <u>Replacing in kind extensively deteriorated or missing parts.</u>

See pages 38 and 39 of the PDF file for *The Secretary of the Interior's Standards for the Treatment of Historic Properties with Guidelines for Preserving, Rehabilitating Restoring & Reconstructing Historic Buildings.*

97. Answer: c
The lead-based paint for a historic building starts to peel, chip, craze, or otherwise comes loose. The following procedure is recommended:
- Remove the lead-paint throughout the building and apply a compatible primer and finish paint (Special license, training and/or protection gear and clothing is often required for removing the hazardous lead-based paint)

The following are NOT recommended:
- Leave the paint in place and do nothing (The lead-based paint for a historic building start to peel, chip, craze, or otherwise comes loose. This is a health and safety concern, and can NOT be left in place. If the existing lead-based paint is in good condition, it can be left in place)
- Use the same kind of lead-based paint to repaint the damaged area (The lead-based paint has been banned in construction since late 1970s.)

98. Answer: a
Cast-in-place concrete beam-and-slab system is the most appropriate structural system for a lab building that is sensitive to vibration.

The following systems are not as good as cast-in-place concrete beam-and-slab system:
- Heavy timber construction with panelized floor
- Lightweight concrete over metal deck over steel joists
- 4" gypsum concrete topping slab over wood deck over wood joists

99. Answer: a and c
The following are appropriate for extinguishing a fire in a cell phone equipment room:
- Dry ice
- Dry chemicals

The following and any other substance containing water are inappropriate for extinguishing a fire in a cell phone equipment room:
- Water
- A combination of water, carbon dioxide and dry chemicals

100. Answer: a
Waterstop is the correct term for the preformed synthetic rubber labeled as letter A in previous image.

The following are incorrect answers:
- Moisture barrier (It is a water-resistant membrane placed under the slab or in the warm side of the walls).
- Control joint (It is a pre-determined concrete joint for concrete to crack along it.)
- Isolation joint (It is a joint that completely separate the concrete or the building.)

101. Answer: c
The image C shows the correct placement of water resistant membrane.

The images A, B and D show the incorrect placement of water resistant membrane:
The water resistant membrane is not placed deep enough, and water and moisture can still penetrate through the retaining wall and gets into the building.

102. Answer: a
The most important factor for locating a supermarket is its proximity to potential clients.

Both a and d are potential correct answers, but answer d (its proximity to high income households) is not as good as answer a (its proximity to potential clients) because the original question does not mention this is a high-end or low-end supermarket, and people from high income households may NOT shop at this supermarket.

The following are incorrect answers:
- the availability of utilities (this is a secondary consideration when compared with others)
- a downtown location (it may not be a good choice because it may not have large enough space for a supermarket, and it may not provide proximity to potential clients)

103. Answer: b
A practice or device designed to keep eroded soil on a construction site, so that it does not wash off and cause water pollution to a nearby water body is called **sediment control**. Sediment control is preventing the *disturbed* soils from entering a nearby water body. Sediment basins and silt fences are examples of sediment control.

Erosion control is the practice of controlling wind or water erosion in land development, agriculture, and construction. Erosion control is *holding* the soils *in place*. It often involves a physical barrier, such as vegetation or rock.

The following are incorrect answers:
- Pollution control
- Defoliation control
- erosion control

104. Answer: a
Please note this question is regarding the *southern* hemisphere.

A roof overhang on which of the *north* façades of a building built in the *southern* hemisphere will provide seasonal adjustment for solar radiation.

If the question is regarding the *northern* hemisphere, then the answer should be b, South (façade).

105. Answer: a and b
The following are the most important factors in the design of residential units because they are important for ensuring each unit can receive sun for at least part of a winter day:
- Orientation
- The heights and locations of adjacent buildings

The following are incorrect answers:
- Bedrooms facing the dominant wind (It is important, but not the most important factors among the four choices).
- Avoiding west facing units (It is important, but not the most important factors among the four choices).

106. Answer: a and c

    The following are correct:
    - **Detention ponds** are also "dry ponds."
    - **Detention ponds** are used to hold stormwater for a short period of time.

    Retention ponds are "wet ponds." They are artificial lakes with vegetation around the perimeter.

107. Answer: c, d and f

    Pay attention to the words "EXCEPT."

    The following are NOT considered an environmental impact issue for site analysis, and therefore the correct answers:
    - Recommended footing design
    - Archeological discoveries
    - Demography

    The following are considered an environmental impact issue for site analysis, and therefore the incorrect answers:
    - Reflections
    - Dominant wind direction
    - Sun and shadow patterns
    - Traffic condition

108. Answer: c

    Restrictive covenants are typically controlled by HOA, or Home Owners' Association.
    The following are incorrect answers:
    - City
    - Contractor
    - EPA

109. Answer: e

    Biophilia means an instinctive bond between human beings and other living systems.

    The following are incorrect answers:
    - Nearsightedness
    - Farsightedness
    - Love at the first sight
    - Human beings' latent desire of being loved

110. Answer: d

Looking at the cost of purchasing and operating a building or product, and the relative savings is called **life cycle costing**.

Life cycle approach: Looking at a product or building through its entire life cycle.

Life cycle assessment (LCA): Use life cycle thinking in environmental issues.

Low impact development (LID): A land development approach mimicking natural systems and managing storm water as close to the source as possible.

Life cycle analysis: a technique to assess environmental impacts associated with all the stages of a product's or life cycle.

Life cycle cost and saving analysis is an invented term used as a distracter to confuse you.

B. Mock Exam Answers and Explanations: Case Study

111. Answer: b
    The program requires an exterior window for the Small Meeting Room. No exterior window is needed for the Stacks, and a view is not required from either the Head Librarian's Office or the Assistant Librarian's Office.

112. Answer: c and d
    Please note that we are looking for the *incorrect* statements. Answers c and d are incorrect statements and therefore the correct answers. There is no information in the program that defines the exit corridor width. Exit doors can pass through a stair in some cases.

    The double doors for the MR swing in the wrong direction. The two doors that open to outside from the two stairs *on the second floor* are not necessary, and are drafting errors.

113. Answer: a and b
    Visual control requirements are not completely met because the Lending Desk/Office has no visual control of the entry to the Lobby or the Children's Reading Room. A window is missing between the Lending Desk/Office and the Lobby. The Workroom does not have the required direct access to Lending Desk/Office. A door is missing between these two rooms.

    The program does not require Toilet Rooms to have windows. The Large Meeting Room does not need to have two exits.

114. Answer:
    Access to a public right of way should be from the front entry and all the other exits. See the four shaded areas on figure 4.5 for acceptable locations.

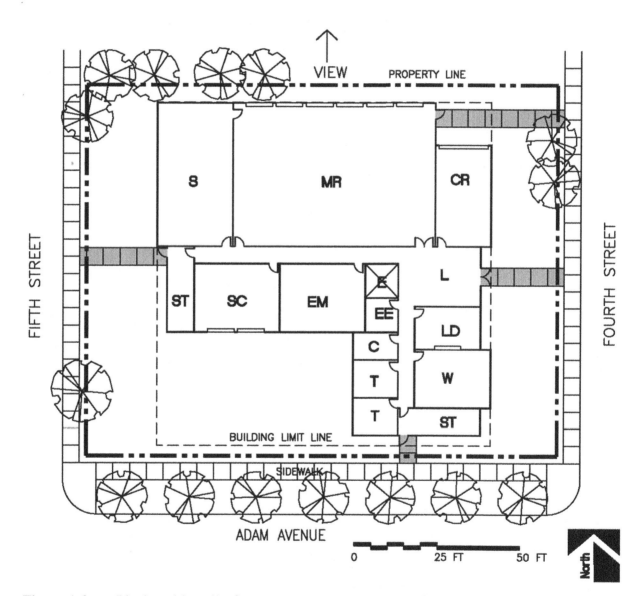

Figure 4.6   Placing sidewalks for access to the public right of the way.

115. Answer: d

Both the Head Librarian's Office and Assistant Librarian's Office need to have direct access to Secretarial Office, so choices b and c are both incorrect because the new 350 sf Meeting Room will block off the direct access.

Answers a and d are both possible solutions, but choice d is better because the second floor can expand to the east, and align with the exterior walls of the first floor to save on construction costs.

116. Answer: b and c

Per the program, Special Collections must be on the first floor, so choice a is incorrect. The Workroom is required to have direct access to Lending Desk/Office, which needs to stay on the first floor in order to have a direct view of Lobby entrance, so choice d is incorrect.

The Elevator Equipment Room and Electrical/Mechanical Room can be moved to the second floor. Ideally the Electrical/Mechanical Room should be placed on the first floor because the room contains heavy equipment, but the program does not prohibit placing it on the second floor. Therefore, choice f will be incorrect, and choices b and c are correct.

117. Answer: c

Concrete and clay tiles may be installed on roofs with slopes of 2:12 and greater, however, roofs with slopes of 4:12 and less must have a *double* layer of underlayment. Choices a and b only list one layer of underlayment, and are therefore incorrect.

Because the roof has no parapet, neither a single-ply TPO roof nor a single-ply PVC roof is a good choice, because the edges will be visible. So choices d and e are incorrect.

Composition roof (asphalt shingles) should be installed on a roof with a pitch of 4:12 or greater, however, most manufactures will warrant their shingles down to a 2:12 pitch if one of the following recommended installation methods is used:
- Installing composition roof with an ice and water membrane throughout the entire roof.
- Installing composition roof with two layers of 15-pound asphalt saturated roofing felt with the lap seams offset from one another and all lap seams embedded in an asphalt cement.

The composition roof will look better than single-ply TPO roof or single-ply PVC roof. So, choice c is correct.

118. Answer: d

Choices a, b and c will make the related rooms exceed the 10% allowable discrepancy of the area required by the program, and will not work.

Choice d will expand the Head Librarian's Office and the Assistant Librarian's Office to align with the exterior wall of Secretarial Office. This modification will not exceed the 10% allowable discrepancy of the areas required by the program.

119. Answer: a

There may be accessibility problems because the doors to the Toilet Rooms do not have a 12" clearance on the push side.

There are enough number of exits to meet all the program exit requirements.

The building is within the building limit line and is *not* too close to the property line. The program does not have any other setback requirements.

The stair is *not* too narrow, and the program does not have specific exit width requirements.

120. Answer: a and b

    Expanding the second floor to the east or to the north to align with the first floor exterior walls will take full advantage of the existing structure, and save on construction costs.

    Expanding the building to the south to take full advantage of the site may create access problems or a dead-end corridor over 20 feet in length. This solution will require a new concrete slab, footing, and exterior walls for the first floor. Savings in construction cost is unlikely.

    Similarly, expanding the building to the north of the Children's Reading Room is unlikely to yield construction cost savings, and will not generate a large enough area to justify the cost. The addition may also block the second exit required for the Main Reading Room.

C. How We Came Up with the PPD Mock Exam Questions

We came up with all the PPD Mock Exam questions based on the ARE 5.0 Handbook, and we developed the Mock Exam based on the *five* weighted sections. See a detailed breakdown in the following tables:

Note: If the text on following tables is too small for you to read, then you can go to our forum, sign up for a free account, and download the FREE 11x17 full size jpeg format files for these tables at:
GeeForum.com

Sections	Expected Number of Items	Actual Number of Items
Section 1: Environmental Conditions & Context (10-16%)	12 to 20	13
• Determine location of building and site improvements based on site analysis (A/E)		4
• Determine sustainable principles to apply to design (A/E)		8
• Determine impact of neighborhood context on the project design (U/A)		1
Section 2: Codes & Regulations (16-22%)	20 to 27	23
• Apply zoning and environmental regulations to site and building design (U/A)		4
• Apply building codes to building design (U/A)		10
• Integrate multiple codes to a project design (A/E)		9
Section 3: Building Systems, Materials, & Assemblies (19-25%)	22 to 30	30
• Determine mechanical, electrical, and plumbing (MEP) systems (A/E)		8
• Determine structural systems (A/E)		13
• Determine special systems such as acoustics, communications, lighting, security, conveying, and fire suppression (A/E)		2
• Determine materials and assemblies to meet programmatic, budgetary, and regulatory requirements (A/E)		7
Section 4: Project Integration of Program & Systems (32-38%)	38 to 46	44
• Determine building configuration (A/E)		5
• Integrate building systems in the project design (A/E)		9
• Integrate program requirements into a project design (A/E)		14
• Integrate environmental and contextual conditions in the project design (A/E)		16
Section 5: Project Costs & Budgeting (8-14%)	9 to 17	10
• Evaluate design alternatives based on the program (A/E)		3
• Perform cost evaluation (A/E)		3
• Evaluate cost considerations during the design process (A/E)		4
Total		120

Appendixes

A. List of Figures

Figure 3.1	Direction of water flow	46
Figure 3.2	Retaining wall section	47
Figure 3.3	Spot elevations	48
Figure 3.4	Types of slopes	49
Figure 3.5	Invert Elevation	50
Figure 3.6	Lock image	51
Figure 3.7	Window type	52
Figure 3.8	Storefront elevation	53
Figure 3.9	Roof detail	54
Figure 3.10	Concrete detail	55
Figure 3.11	Brick pattern	56
Figure 3.12	Steel shapes	57
Figure 3.13	Chart	58
Figure 3.14	Air-Conditioning Distribution System	59
Figure 3.15	Plumbing Drawing	60
Figure 3.16	Refrigeration Flow Diagram	61
Figure 3.17	Switch Wiring Diagram	62
Figure 3.18	Machines and Appliances	63
Figure 3.19	Electrical Symbols	64

Figure 3.20	Wiring Diagram	64
Figure 3.21	Orientation of a Classroom Building	66
Figure 3.22	A truss	67
Figure 3.23	A building's floor plan	68
Figure 3.24	Beam to column connections	68
Figure 3.25	Diagram for a frame	69
Figure 3.26	Diagram for seismic forces	69
Figure 3.27	Diagram for a rigid frame	70
Figure 3.28	The moment of inertia for a composite section	70
Figure 3.29	A simple truss	71
Figure 3.30	Selecting a correct diagram of wind pressure	72
Figure 3.31	South Elevation of a Building	77
Figure 3.32	Brick joints	82
Figure 3.33	The correct term for the preformed synthetic rubber	84
Figure 3.34	The placement of water resistant membrane	85
Figure 3.35	Site Plan	90
Figure 3.36	Bubble diagram for the first floor (not to scale)	91
Figure 3.37	Bubble diagram for the second floor (not to scale)	92
Figure 3.38	Block diagram for the first floor	93
Figure 3.39	Block diagram for the second floor	94
Figure 4.1	A Simple Psychrometric Chart	103
Figure 4.2	Air-Conditioning Distribution System: A Variable Flow System	105
Figure 4.3	Refrigeration Flow Diagram	106

Figure 4.4   A free body diagram for Joint A 111

Figure 4.5   A free body diagram for Joint B 112

Figure 4.6   Placing sidewalks for access to the public right of the way 135

B. Official reference materials suggested by NCARB

1. Resources Available While Testing
Tips:
- *You need to read through these pages several times and become very familiar with them to save time in the real ARE exams.*

United States. American Institute of Steel Construction, Inc. *Steel Construction Manual*; 14th edition. Chicago, Illinois, 2011.

Beam Diagrams and Formulas:
- Simple Beam: Diagrams and Formulas - Conditions 1-3, page 3-213; Conditions 4-6, page 3-214; Conditions 7-9, page 3-215
- Beam Fixed at Both Ends: Diagrams and Formulas - Conditions 15-17, page 3-218
- Beam Overhanging One Support: Diagrams and Formulas - Conditions 24-28, pages 3-221 & 222

Dimensions and Properties:
- W Shapes 44 thru 27: Dimensions and Properties, pages 1-12 thru 17
- W Shapes 24 thru W14x145: Dimensions and Properties, pages 1-18 thru 23
- W Shapes 14x132 thru W4: Dimensions and Properties, pages 1-24 thru 29
- C Shapes: Dimensions and Properties, pages 1-36 & 37
- Angles: Properties, pages 1-42 thru 49
- Rectangular HSS: Dimensions and Properties, pages 1-74 thru 91
- Square HSS: Dimensions and Properties, pages 1-92 thru 95
- Round HSS: Dimensions and Properties, pages 1-96 thru 100

United States. International Code Council, Inc. *2012 International Building Code.* Country Club Hills, Illinois, 2011.

Live and Concentrated Loads:
- Uniform and Concentrated Loads: IBC Table 1607.1, pages 340-341

2. Typical Beam Nomenclature

The following typical beam nomenclature is excerpted from:
United States. American Institute of Steel Construction, Inc. *Steel Construction Manual*; 14th edition. Chicago, Illinois, 2011.

E	Modulus of Elasticity of steel at 29,000 ksi	V_2	Vertical shear at right reaction point, or to left of intermediate reaction of beam, kips
I	Moment of Inertia of beam, in^4	V_3	Vertical shear at right reaction point, or to right of intermediate reaction of beam, kips
L	Total length of beam between reaction point, ft	V_x	Vertical shear at distance x from end of beam, kips
M_{max}	Maximum moment, kip-in	W	Total load on beam, kips
M_1	Maximum moment in left section of beam, kip-in	A	Measured distance along beam, in
M_2	Maximum moment in right section of beam, kip-in	B	Measured distance along beam which may be greater or less than a, in
M_3	Maximum positive moment in beam with combined end moment conditions, kip-in	L	Total length of beam between reaction points, in
M_x	Maximum at distance x from end of beam, kip-in	W	Uniformly distributed load per unit of length, kips/in
P	Concentrated load, kips	w_1	Uniformly distributed load per unit of length nearest left reaction, kips/in
P_1	Concentrated load nearest left reaction, kips	w_2	Uniformly distributed load per unit of length nearest right reaction and of different magnitude than w1, kips/in
P_2	Concentrated load nearest right reaction and of different magnitude than P_1, kips	X	Any distance measured along beam from left reaction, in
R	End beam reaction for any condition of symmetrical loading, kips	x_1	Any distance measured along overhang section of beam from nearest reaction point, in
R_1	Left end beam reaction, kips	Δ_{max}	Maximum deflection, in
R_2	Right end or intermediate beam reaction, kips	Δa	Deflection at point of load, in
R_3	Right end beam reaction, kips	Δx	Deflection at point x distance from left reaction, in
V	Maximum vertical shear for any condition of symmetrical loading, kips	Δx_1	Deflection of overhang section of beam at any distance from nearest reaction point, in

| V_1 | Maximum vertical shear in left section of beam, kips | | |

3. Formulas Available While Testing

Tips:
- *These formulas and references will be available during the real exam. You should read through them a few times before the exam to become familiar with them. This will save you a lot of time during the real exam, and will help you solve structural calculations and other problems.*

Structural:
Flexural stress at extreme fiber
$$f = \frac{Mc}{I} = \frac{M}{S}$$

Flexural stress at any fiber
$$f = \frac{My}{I}$$

where y = distance from neutral axis to fiber

Average vertical shear
$$v = \frac{V}{A} = \frac{V}{dt}$$
for beams and girders

Horizontal shearing stress at any section A-A
$$v = \frac{VQ}{Ib}$$
where Q = statical moment about the neutral axis of the entire section of that portion of the cross-section lying outside of section A-A
b = width at section A-A

Electrical
$$Foot-candles = \frac{lumens}{area\ in\ ft^2}$$

$$Foot-candles = \frac{(lamp\ lumens)\ x\ (lamps\ per\ fixture)\ x\ (number\ of\ fixtures)\ x\ (CU)\ x\ (LLF)}{area\ in\ ft^2}$$

$$Number\ of\ luminaires = \frac{(foot-candles)\ x\ (floor\ area)}{(lumens)\ x\ (CU)\ x\ (LLF)}$$

where CU = coefficient of utilization
LLF = Light Loss Factor

$$DF_{AV} = 0.2x \frac{window\ area}{floor\ area}$$

for spaces with sidelighting or toplighting with vertical monitors

watts = volts x amperes x power factor
for AC circuits only

Demand charge = maximum power demand x demand tariff

Plumbing

$1\ psi = 2.31\ feet\ of\ water$

$1\ cubic\ foot = 7.5\ U.S.\ gallons$

HVAC

$$\frac{BTU}{year} = peak\ heat\ loss\ x\ \frac{full - load\ hours}{year}$$

$$\frac{\$}{year} = \frac{BTU}{year} \times \frac{fuel\ cost}{fuel\ heat\ value} \times efficiency$$

$BTU/h = (cfm)\ x\ (1.08)\ x\ (\Delta T)$

$1\ kWh = 3,400\ BTU/h$

$1\ ton\ of\ air\ conditioning = 12,000\ BTU/h$

$BTU/h = (U)\ x\ (A)\ x\ (T_d)$ where Td is the difference between indoor and outdoor temperatures

$U = 1/R_t$

$$U_o = \frac{(U_w \times A_w) + (U_{op} \times A_{op})}{Ao}$$
where o = total wall, w = window, and op = opaque wall

$$U_o = \frac{(U_R \times A_R) + (U_S \times A_S)}{Ao}$$
where o = total roof, R = roof, and S = skylight

$$R = x/k$$
where x = thickness of material in inches

$$Heat\ required = \frac{BTU/h}{temperature\ differential} \times (24\ hours) \times (DD\ °F)$$
where DD = degree days

Acoustics

$$\lambda = \frac{c}{f}$$
where λ = wavelength of sound (ft)
c = velocity of sound (fps)
f = frequency of sound (Hz)

$$a = SAC \times S$$
where a = Absorption of a material used in space (sabins)
SAC = Sound Absorption Coefficient of the material
S = Exposed surface area of the material (ft^2)

$$A = \Sigma a$$
Where A = Total sound absorption of a room (sabins)
$\Sigma a = (S_1 \times SAC_1) + (S_2 \times SAC_2) + ...$

$$T = 0.05 \times \frac{V}{A}$$
where T = Reverberation time (seconds)
V = Volume of space (ft^3)

$$NRC = average\ SAC\ for\ frequency\ bands\ 250, 500, 1000, and\ 2000\ Hz$$

4. Common Abbreviations

Tips:
- *You need to read through these common abbreviations several times and become very familiar with them to save time in the real ARE exams.*

Professional Organizations, Societies, and Agencies

American Concrete Institute	ACI
American Institute of Architects	AIA
American Institute of Steel Construction	AISC
American National Standards Institute	ANSI
American Society for Testing and Materials	ASTM
American Society of Civil Engineers	ASCE
American Society of Heating, Refrigerating, and Air-Conditioning Engineers	ASHRAE
American Society of Mechanical Engineers	ASME
American Society of Plumbing Engineers	ASPE
Architectural Woodwork Institute	AWI
Construction Specifications Institute	CSI
Department of Housing and Urban Development	HUD
Environmental Protection Agency	EPA
Federal Emergency Management Agency	FEMA
National Fire Protection Association	NFPA
Occupational Safety and Health Administration	OSHA
U.S. Green Building Council	USGBC

Tips:
- *You need to look through the following codes and regulations & AIA contract documents several times and become very familiar with them to save time in the real ARE exams. Read some of the important sections in details.*

AIA Contract Documents

A101-2007, Standard Form of Agreement Between Owner and Contractor - Stipulated Sum	A101
A201-2007, General Conditions of the Contract for Construction	A201
A305-1986, Contractor's Qualification Statement	A305
A701-1997, Instructions to Bidders	A701
B101-2007, Standard Form of Agreement Between Owner and Architect	B101
C401-2007, Standard Form of Agreement Between Architect and Consultant	C401
G701-2001, Change Order	G701
G702-1992, Application and Certificate for Payment	G702
G703-1992, Continuation Sheet	G703
G704-2000, Certificate of Substantial Completion	G704

Codes and Regulations

ADA Standards for Accessible Design	ADA

International Code Council	ICC
International Building Code	IBC
International Energy Conservation Code	IECC
International Existing Building Code	IEBC
International Mechanical Code	IMC
International Plumbing Code	IPC
International Residential Code	IRC
Leadership in Energy and Environmental Design	LEED
National Electrical Code	NEC

Commonly Used Terms

Air Handling Unit	AHU
Authority Having Jurisdiction	AHJ
Building Information Modeling	BIM
Concrete Masonry Unit	CMU
Contract Administration	CA
Construction Document	CD
Dead Load	DL
Design Development	DD
Exterior Insulation and Finish System	EIFS
Furniture, Furnishings & Equipment	FF&E
Floor Area Ratio	FAR
Heating, Ventilating, and Air Conditioning	HVAC
Insulating Glass Unit	IGU
Indoor Air Quality	IAQ
Indoor Environmental Quality	IEQ
Live Load	LL
Material Safety Data Sheets	MSDS
Photovoltaic	PV
Reflected Ceiling Plan	RCP
Schematic Design	SD
Variable Air Volume	VAV
Volatile Organic Compound	VOC
British Thermal Unit	btu
Cubic Feet per Minute	cfm
Cubic Feet per Second	cfs
Cubic Foot	cu. ft. ft^3
Cubic Inch	cu. in. in^3
Cubic Yard	cu. yd. yd^3
Decibel	dB
Foot	ft
Foot-candle	fc
Gross Square Feet	gsf

Impact Insulation Class	IIC
Inch	in
Net Square Feet	nsf
Noise Reduction Coefficient	NRC
Pound	lb
Pounds per Linear Foot	plf
Pounds per Square Foot	psf
Pounds per Square Inch	psi
Sound Transmission Class	STC
Square Foot	sq. ft.
	sf
	ft^2
Square Inch	sq. in.
	in^2
Square Yard	sq. yd.

5. **General NCARB reference materials for ARE:**

Per NCARB, all candidates should become familiar with the latest version of the following codes:

International Code Council, Inc. (ICC)
International Building Code
International Mechanical Code
International Plumbing Code

National Fire Protection Association (NFPA)
Life Safety Code (NFPA 101)
National Electrical Code (NFPA 70)

National Research Council of Canada
National Building Code of Canada
National Plumbing Code of Canada
National Fire Code of Canada

American Institute of Architects
AIA Documents - 2007

6. Official NCARB reference materials matrix

Per NCARB, all candidates should become familiar with the latest version of the following:

Reference	PcM	PjM	PA	PPD	PDD	CE
2009 ASHRAE Handbook: Fundamentals, I-P Edition. ASHRAE, 2009				✓	✓	
2010 ADA Standards for Accessible Design U.S. Department of Justice, 2010			✓	✓		
The American Institute of Architects Official Guide to the 2007 AIA Contract Documents. The American Institute of Architects. John Wiley & Sons, 2009	✓	✓				✓
The Architect's Guide to Small Firm Management: Making Chaos Work for Your Small Firm. Rena M. Klein, FAIA. The American Institute of Architects. John Wiley & Sons, 2010	✓					
The Architect's Handbook of Professional Practice The American Institute of Architects John Wiley & Sons, latest edition	✓	✓		✓	✓	
Architectural Acoustics. M. David Egan. J. Ross Publishing, 2007. Reprint. Original publication McGraw Hill, 1988				✓	✓	
Architectural Graphic Standards The American Institute of Architects John Wiley & Sons, latest edition				✓	✓	
Architectural Graphic Standards for Residential Construction The American Institute of Architects John Wiley & Sons, latest edition				✓	✓	
BIM and Integrated Design: Strategies for Architectural Practice Randy Deutsch, AIA, LEED-AP The American Institute of Architects John Wiley & Sons, 2011	✓	✓				
Building Codes Illustrated: A Guide to Understanding the 2012 International Building Code. Francis D.K. Ching and Steven R. Winkel, FAIA, PE. John Wiley & Sons, 2012			✓	✓	✓	
Building Construction Illustrated Francis D. K. Ching John Wiley & Sons, latest edition				✓	✓	
Building Structures James Ambrose and Patrick Tripeny John Wiley & Sons, 3rd edition, 2012				✓	✓	

Reference	PcM	PjM	PA	PPD	PDD	CE
Code of Ethics and Professional Conduct AIA Office of General Counsel. The American Institute of Architects, latest edition	■					
CSI MasterFormat. The Construction Specifications Institute. 2004 edition						■
Dictionary of Architecture and Construction. Cyril M. Harris. McGraw-Hill, 4th edition, 2006				■	■	■
Fundamentals of Building Construction: Materials and Methods Edward Allen and Joseph Iano John Wiley & Sons, latest edition				■	■	
Heating, Cooling, Lighting: Sustainable Design Methods for Architects. Norbert Lechner. John Wiley & Sons, 3rd edition, 2008				■	■	
The HOK Guidebook to Sustainable Design Sandra F. Mendler, William Odell, and Mary Ann Lazarus John Wiley & Sons, 2nd edition, 2006				■	■	
ICC A117.1-2009 Accessible and Usable Buildings and Facilities International Code Council, 2010			■			
International Building Code (2012) International Code Council, 2011			■	■		
Landscape Architectural Graphic Standards. Leonard J. Hopper, Editor. John Wiley & Sons, 2007				■		
Law for Architects: What You Need to Know. Robert F. Herrmann and the Attorneys at Menaker & Herrmann LLP. W. W. Norton, 2012	■					
Mechanical & Electrical Equipment for Buildings. Walter T. Grondzik, Alison G. Kwok, Benjamin Stein, and John S. Reynolds, Editors. John Wiley & Sons, latest edition				■	■	
Mechanical and Electrical Systems in Buildings. Richard R. Janis and William K. Y. Tao. Prentice Hall, latest edition.				■	■	
Minimum Design Loads for Buildings and Other Structures (7-10) American Society of Civil Engineers, 2013					■	
Olin's Construction Principles, Materials, and Methods. H. Leslie Simmons. John Wiley & Sons, latest edition				■	■	

Reference	PcM	PjM	PA	PPD	PDD	CE
Planning and Urban Design Standards American Planning Association John Wiley & Sons, 2006			■			
Plumbing, Electricity, Acoustics: Sustainable Design Methods for Architecture. Norbert Lechner. John Wiley & Sons, 2012				■	■	
Problem Seeking: An Architectural Programming Primer William M. Peña and Steven A. Parshall John Wiley & Sons, latest edition			■			
Professional Practice: A Guide to Turning Designs into Buildings. Paul Segal, FAIA. W. W. Norton, 2006	■	■	■			
The Professional Practice of Architectural Working Drawings. Osamu A. Wakita, Nagy R. Bakhoum, and Richard M. Linde. John Wiley & Sons, 4th edition, 2012					■	
The Project Resource Manual: CSI Manual of Practice. The Construction Specifications Institute. McGraw-Hill, 5th edition, 2005						■
Rules of Conduct National Council of Architectural Registration Boards, latest edition	■					
Simplified Engineering for Architects and Builders James Ambrose and Patrick Tripeny John Wiley & Sons, latest edition					■	
Site Engineering for Landscape Architects Steven Strom, Kurt Nathan, and Jake Woland, Editors. John Wiley & Sons, 2013				■		
Site Planning and Design Handbook Thomas H. Russ McGraw-Hill, 2nd edition, 2009			■	■		
Space Planning Basics Mark Karlen and Rob Fleming John Wiley & Sons, latest edition			■			
Steel Construction Manual American Institute of Steel Construction Ingram, latest edition					■	
Structural Design: A Practical Guide for Architects James R. Underwood and Michele Chiuini John Wiley & Sons, 2nd edition, 2007				■	■	

Reference	PcM	PjM	PA	PPD	PDD	CE
Structures Daniel Schodek and Martin Bechthold Pearson/Prentice Hall, latest edition				■	■	
Sun, Wind, and Light: Architectural Design Strategies G.Z. Brown and Mark DeKay John Wiley & Sons, 2nd edition, 2001				■		
Sustainable Construction: Green Building Design and Delivery Charles J. Kibert. John Wiley & Sons, 2005				■		
Time-Saver Standards for Architectural Design: Technical Data for Professional Practice Donald Watson and Michael J. Crosbie, Editors. McGraw-Hill, 8th edition, 2004				■	■	
A Visual Dictionary of Architecture Francis D.K. Ching John Wiley & Sons, latest edition				■		

The following AIA Contract Documents have content covered in the of ARE 5.0 exams. Candidates can access them for free through their NCARB Record.

Conventional Family	PcM	PjM	PA	PPD	PDD	CE
A101-2007, Standard Form of Agreement Between Owner and Contractor where the basis of payment is a Stipulated Sum						■
A201-2007, General Conditions of the Contract for Construction		■				■
A701-1997, Instructions to Bidders		■				
B101-2007, Standard Form of Agreement Between Owner and Architect	■					
C401-2007, Standard Form of Agreement Between Architect and Consultant	■					
Contract Administration and Project Management Forms	PcM	PjM	PA	PPD	PDD	CE
A305-1986, Contractor's Qualification Statement						■
G701-2001, Change Order						■
G702-1992, Application and Certificate for Payment						■
G703-1992, Continuation Sheet						■
G704-2000, Certificate of Substantial Completion						■

The following are some extra study materials if you have some additional time and want to learn more. If you are tight on time, you can simply look through them and focus on the sections that cover your weakness:

ACI Code 318-05 (Building Code Requirements for Reinforced Concrete)
American Concrete Institute, 2005

OR
CAN/CSA-A23.1-94 (Concrete Materials and Methods of Concrete Construction) and CAN/CSA-A23.3-94 (Design of Concrete Structures for Buildings)
Canadian Standards Association

Design Value for Wood Construction
American Wood Council, 2005

Elementary Structures for Architects and Builders, Fourth Edition
Ronald E. Shaeffer
Prentice Hall, 2006

Introduction to Wood Design
Canadian Wood Council, 2005

Manual of Steel Construction: Allowable Stress Design; 9th Edition.
American Institute of Steel Construction, Inc. Chicago, Illinois, 1989

National Building Code of Canada, 2005
Parts 1, 3, 4, 9; Appendix A
Supplement
Chapters 1, 2, 4; Commentaries A, D, F, H, I

NEHRP (National Earthquake Hazards Reduction Program) Recommended Provisions for Seismic Regulations for New Buildings and Other Structures Parts 1 and 2
FEMA 2003

Simplified Building Design for Wind and Earthquake Forces
James Ambrose and Dimitry Vergun
John Wiley & Sons, 1997

Simplified Design of Concrete Structures,
Eighth Edition
James Ambrose, Patrick Tripeny
John Wiley & Sons, 2007

Simplified Design of Masonry Structures
James Ambrose
John Wiley & Sons, 1997

Simplified Design of Steel Structures, Eighth Edition
James Ambrose, Patrick Tripeny
John Wiley & Sons, 2007

Simplified Design of Wood Structures, Fifth Edition
James Ambrose
John Wiley & Sons, 2009

Simplified Mechanics and Strength of Materials, Fifth Edition
Harry Parker and James Ambrose
John Wiley & Sons, 2002

Standard Specifications Load Tables &Weight Tables for Steel Joists and Joist Girders
Steel Joist Institute, latest edition

Steel Construction Manual, Latest edition
American Institute of Steel Construction, 2006

OR
Handbook of Steel Construction, Latest edition; and *CAN/CSA-S16-01 and CISC Commentary*
Canadian Institute of Steel Construction

Steel Deck Institute Tables
Steel Deck Institute

OR
LSD Steel Deck Tables
Caradon Metal Building Products

Structural Concepts and Systems for Architects and Engineers, Second Edition
T.Y. Lin and Sidney D. Stotesbury
Van Nostrand Reinhold, 1988

Structural Design: A Practical Guide for Architects
James Underwood and Michele Chiuini
John Wiley & Sons, latest edition

Structure in Architecture: The Building of Buildings
Mario Salvadori with Robert Heller
Prentice-Hall, 1986

Understanding Structures
Fuller Moore

McGraw-Hill, 1999

Wood Design Manual and *CAN/CSA-086.1-94 and Commentary*
Canadian Wood Council

C. Other reference materials

Chen, Gang. *Building Construction: Project Management, Construction Administration, Drawings, Specs, Detailing Tips, Schedules, Checklists, and Secrets Others Don't Tell You (Architectural Practice Simplified, 2nd edition)*. ArchiteG, Inc., A good introduction to the architectural practice and construction documents and service, including discussions of MasterSpec format and specification sections.

Chen, Gang. ***LEED v4 Green Associate Exam Guide (LEED GA):*** *Comprehensive Study Materials, Sample Questions, Mock Exam, Green Building LEED Certification, and Sustainability*, Book 2, LEED Exam Guide series, ArchiteG.com, the latest edition. ArchiteG, Inc. Latest Edition. This is a very comprehensive and concise book on the LEED Green Associate Exam. Some readers have passed the LEED Green Associate Exam by studying this book for 10 hours.

Ching, Francis. *Architecture: Form, Space, & Order.* Wiley, latest edition. It is one of the best architectural books that you can have. I still flip through it every now and then. It is a great book for inspiration.

Frampton, Kenneth. *Modern Architecture: A Critical History*. Thames and Hudson, London, latest edition. A valuable resource for architectural history.

Jarzombek, Mark M. (Author), Vikramaditya Prakash (Author), Francis D. K. Ching (Editor). *A Global History of Architecture*. Wiley, latest edition. A valuable and comprehensive resource for architectural history with 1000 b & w photos, 50 color photos, and 1500 b & w illustrations. It doesn't limit the topic on a Western perspective, but rather through a global vision.

Trachtenberg, Marvin and Isabelle Hyman. *Architecture: From Pre-history to Post-Modernism*. Prentice Hall, Englewood Cliffs, NJ latest edition. A valuable and comprehensive resource for architectural history.

D. Some Important Information about Architects and the Profession of Architecture

What Architects Do?

Architects plan and design houses, factories, office buildings, and other structures.

Duties
Architects typically do the following:

- Meet with clients to determine objectives and requirements for structures
- Give preliminary estimates on cost and construction time
- Prepare structure specifications
- Direct workers who prepare drawings and documents
- Prepare scaled drawings, either with computer software or by hand
- Prepare contract documents for building contractors
- Manage construction contracts
- Visit worksites to ensure that construction adheres to architectural plans
- Seek new work by marketing and giving presentations

People need places to live, work, play, learn, shop, and eat. Architects are responsible for designing these places. They work on public or private projects and design both indoor and outdoor spaces. Architects can be commissioned to design anything from a single room to an entire complex of buildings.

Architects discuss the objectives, requirements, and budget of a project with clients. In some cases, architects provide various predesign services, such as feasibility and environmental impact studies, site selection, cost analyses, and design requirements.

Architects develop final construction plans after discussing and agreeing on the initial proposal with clients. These plans show the building's appearance and details of its construction. Accompanying these plans are drawings of the structural system; air-conditioning, heating, and ventilating systems; electrical systems; communications systems; and plumbing. Sometimes, landscape plans are included as well. In developing designs, architects must follow state and local building codes, zoning laws, fire regulations, and other ordinances, such as those requiring easy access to buildings for people who are disabled.

Computer-aided design and drafting (CADD) and building information modeling (BIM) have replaced traditional drafting paper and pencil as the most common methods for creating designs and construction drawings. However, hand-drawing skills are still required, especially during the conceptual stages of a project and when an architect is at a construction site.

As construction continues, architects may visit building sites to ensure that contractors follow the design, adhere to the schedule, use the specified materials, and meet work-quality standards. The job is not complete until all construction is finished, required tests are conducted, and construction costs are paid.

Architects may also help clients get construction bids, select contractors, and negotiate construction contracts.

Architects often collaborate with workers in related occupations, such as civil engineers, urban and regional planners, drafters, interior designers, and landscape architects.

Work Environment
Although architects usually work in an office, they must also travel to construction sites.

Architects held about 112,600 jobs in 2014, with 69 percent employed in architectural, engineering, and related services. About 1 in 5 were self-employed.

Architects spend much of their time in offices, where they meet with clients, develop reports and drawings, and work with other architects and engineers. They also visit construction sites to ensure clients' objectives are met and to review the progress of projects. Some architects work from home offices.

Work Schedules
Most architects work full time and many work additional hours, especially when facing deadlines. Self-employed architects may have more flexible work schedules.

How to Become an Architect
There are typically three main steps to becoming a licensed architect: completing a professional degree in architecture, gaining relevant experience through a paid internship, and passing the Architect Registration Examination.

Education
In all states, earning a professional degree in architecture is typically the first step to becoming an architect. Most architects earn their professional degree through a 5-year Bachelor of Architecture degree program, intended for students with no previous architectural training. Many earn a master's degree in architecture, which can take 1 to 5 years in addition to the time spent earning a bachelor's degree. The amount of time required depends on the extent of the student's previous education and training in architecture.

A typical bachelor's degree program includes courses in architectural history and theory, building design with an emphasis on computer-aided design and drafting (CADD), structures, construction methods, professional practices, math, physical sciences, and liberal arts. Central to most architectural programs is the design studio, where students apply the skills and concepts learned in the classroom to create drawings and three-dimensional models of their designs.

Currently, 34 states require that architects hold a professional degree in architecture from one of the 123 schools of architecture accredited by the National Architectural Accrediting Board (NAAB). State licensing requirements can be found at the National Council of Architectural Registration Boards (NCARB). In the states that do not have that requirement, applicants can become licensed with 8 to 13 years of related work experience in addition to a high school diploma. However, most architects in these states still obtain a professional degree in architecture.

Training
All state architectural registration boards require architecture graduates to complete a lengthy paid internship—generally 3 years of experience—before they may sit for the Architect Registration Examination. Most new graduates complete their training period by working at architectural firms through the Intern Development Program (IDP), a program run by NCARB that guides students through the internship process. Some states allow a portion of the training to occur in the offices of employers in related careers, such as engineers and general contractors. Architecture students who complete internships while still in school can count some of that time toward the 3-year training period.

Interns in architectural firms may help design part of a project. They may help prepare architectural documents and drawings, build models, and prepare construction drawings on CADD. Interns may also research building codes and write specifications for building materials, installation criteria, the quality of finishes, and other related details. Licensed architects will take the documents that interns produce, make edits to them, finalize plans, and then sign and seal the documents.

Licenses, Certifications, and Registrations
All states and the District of Columbia require architects to be licensed. Licensing requirements typically include completing a professional degree in architecture, gaining relevant experience through a paid internship, and passing the Architect Registration Examination.

Most states also require some form of continuing education to keep a license, and some additional states are expected to adopt mandatory continuing education. Requirements vary by state but usually involve additional education through workshops, university classes, conferences, self-study courses, or other sources.

A growing number of architects voluntarily seek certification from NCARB. This certification makes it easier to become licensed across states, because it is the primary requirement for reciprocity of licensing among state boards that are NCARB members. In 2014, approximately one-third of all licensed architects had the certification.

Advancement
After many years of work experience, some architects advance to become architectural and engineering managers. These managers typically coordinate the activities of employees and may work on larger construction projects.

Important Qualities
Analytical skills. Architects must understand the content of designs and the context in which they were created. For example, architects must understand the locations of mechanical systems and how those systems affect building operations.

Communication skills. Architects share their ideas, both in oral presentations and in writing, with clients, other architects, and workers who help prepare drawings. Many also give presentations to explain their ideas and designs.

Creativity. Architects design the overall look of houses, buildings, and other structures. Therefore, the final product should be attractive and functional.

Organizational skills. Architects often manage contracts. Therefore, they must keep records related to the details of a project, including total cost, materials used, and progress.

Technical skills. Architects need to use CADD technology to create plans as part of building information modeling (BIM).

Visualization skills. Architects must be able to see how the parts of a structure relate to each other. They also must be able to visualize how the overall building will look once completed.

Pay

The median annual wage for architects was $76,100 in May 2015. The median wage is the wage at which half the workers in an occupation earned more than that amount and half earned less. The lowest 10 percent earned less than $46,080, and the highest 10 percent earned more than $125,520.

Some firms pay tuition and fees toward continuing education requirements for their employees. Most architects work full time and many work additional hours, especially when facing deadlines. Self-employed architects may have more flexible work hours.

Job Outlook

Employment of architects is projected to grow 7 percent from 2014 to 2024, about as fast as the average for all occupations.

Architects will be needed to make plans and designs for the construction and renovation of homes, offices, retail stores, and other structures. Many school districts and universities are expected to build new facilities or renovate existing ones. In addition, demand is expected for more healthcare facilities as the baby-boomer population ages and as more individuals use healthcare services. The construction of new retail establishments may also require more architects.

Demand is projected for architects with a knowledge of "green design," also called sustainable design. Sustainable design emphasizes the efficient use of resources, such as energy and water conservation; waste and pollution reduction; and environmentally friendly design, specifications, and materials. Rising energy costs and increased concern about the environment have led to many new buildings being built with more sustainable designs.

The use of CADD and, more recently, BIM, has made architects more productive. These technologies have allowed architects to do more work without the help of drafters while making it easier to share the work with engineers, contractors, and clients.

Job Prospects
With a high number of students graduating with degrees in architecture, very strong competition for internships and jobs is expected. Competition for jobs will be especially strong at the most prestigious architectural firms. Those with up-to-date technical skills—including a strong grasp of CADD and BIM—and experience in sustainable design will have an advantage.

Employment of architects is strongly tied to the activity of the construction industry. Therefore, these workers may experience periods of unemployment when there is a slowdown in requests for new projects or when the overall level of construction falls.

State & Area Data
Occupational Employment Statistics (OES)
The Occupational Employment Statistics (OES) program produces employment and wage estimates annually for over 800 occupations. These estimates are available for the nation as a whole, for individual states, and for metropolitan and nonmetropolitan areas. The link(s) below go to OES data maps for employment and wages by state and area.
https://www.bls.gov/oes/current/oes171011.htm#st

Projections Central
Occupational employment projections are developed for all states by Labor Market Information (LMI) or individual state Employment Projections offices. All state projections data are available at www.projectionscentral.com. Information on this site allows projected employment growth for an occupation to be compared among states or to be compared within one state. In addition, states may produce projections for areas; there are links to each state's websites where these data may be retrieved.

Career InfoNet
America's Career InfoNet includes hundreds of occupational profiles with data available by state and metro area. There are links in the left-hand side menu to compare occupational employment by state and occupational wages by local area or metro area. There is also a salary info tool to search for wages by zip code.

Related Occupations
Architects design buildings and related structures. Construction managers, like architects, also plan and coordinate activities concerned with the construction and maintenance of buildings and facilities. Others who engage in similar work are landscape architects, civil engineers, urban and regional planners, and designers, including interior designers, commercial and industrial designers, and graphic designers.
Sources of Additional Information

Disclaimer:
Links to non-BLS Internet sites are provided for your convenience and do not constitute an endorsement.

Information about education and careers in architecture can be obtained from:
- The American Institute of Architects, 1735 New York Ave. NW., Washington, DC 20006. Internet: http://www.aia.org
- National Architectural Accrediting Board: http://www.naab.org/
- National Council of Architectural Registration Boards, Suite 1100K, 1801 K St. NW., Washington, D.C. 20006. Internet: http://www.ncarb.org
  OOH ONET Codes 17-1011.00"

Source: Bureau of Labor Statistics, U.S. Department of Labor, *Occupational Outlook Handbook, 2016-17 Edition*, Architects, on the Internet at https://www.bls.gov/ooh/architecture-and-engineering/architects.htm (visited **January 26, 2017**).

Publish Date: Thursday, December 17, 2015

Note: Please check the website above for the latest information.

E. AIA Compensation Survey

Every 3 years, AIA publishes a Compensation Survey for various positions at architectural firms across the country. It is a good idea to find out the salary before you make the final decision to become an architect. If you are already an architect, it is also a good idea to determine if you are underpaid or overpaid.

See following link for some sample pages for the 2015 AIA Compensation Survey:

https://www.aia.org/resources/8066-aia-compensation-report

F. So ... You would Like to Study Architecture

To study architecture, you need to learn how to draft, how to understand and organize spaces and the interactions between interior and exterior spaces, how to do design, and how to communicate effectively. You also need to understand the history of architecture.

As an architect, a leader for a team of various design professionals, you not only need to know architecture, but also need to understand enough of your consultants' work to be able to coordinate them. Your consultants include soils and civil engineers, landscape architects, structural, electrical, mechanical, and plumbing engineers, interior designers, sign consultants, etc.

There are two major career paths for you in architecture: practice as an architect or teach in colleges or universities. The earlier you determine which path you are going to take, the more likely you will be successful at an early age. Some famous and well-respected architects, like my USC alumnus Frank Gehry, have combined the two paths successfully. They teach at the universities and have their own architectural practice. Even as a college or university professor, people respect you more if you have actual working experience and have some built projects. If you only teach in colleges or universities but have no actual working experience and have no built projects, people will consider you as a "paper" architect, and they are not likely to take you seriously, because they will think you probably do not know how to put a real building together.

In the U.S., if you want to practice architecture, you need to obtain an architect's license. It requires a combination of passing scores on the Architectural Registration Exam (ARE) and 8 years of education and/or qualified working experience, including at least 1 year of working experience in the U.S. Your working experience needs to be under the supervision of a licensed architect to be counted as qualified working experience for your architect's license.

If you work for a landscape architect or civil engineer or structural engineer, some states' architectural licensing boards will count your experience at a discounted rate for the qualification of your architect's license. For example, 2 years of experience working for a civil engineer may be counted as 1 year of qualified experience for your architect's license. You need to contact your state's architectural licensing board for specific licensing requirements for your state.

If you want to teach in colleges or universities, you probably want to obtain a master's degree or a Ph.D. It is not very common for people in the architectural field to have a Ph.D. One reason is that there are few Ph.D. programs for architecture. Another reason is that architecture is considered a profession and requires a license. Many people think an architect's license is more important than a Ph.D. degree. In many states, you need to have an architect's license to even use the title "architect," or the terms "architectural" or "architecture" to advertise your service. You cannot call yourself an architect if you do not have an architect's license, even if you have a Ph.D. in architecture. Violation of these rules brings punishment.

To become a tenured professor, you need to have a certain number of publications and pass the evaluation for the tenure position. Publications are very important for tenure track positions. Some people say for the tenured track positions in universities and colleges, it is "publish or perish."

The American Institute of Architects (AIA) is the national organization for the architectural profession. Membership is voluntary. There are different levels of AIA membership. Only licensed architects can be (full) AIA members. If you are an architectural student or an intern but not a licensed architect yet, you can join as an associate AIA member. Contact AIA for detailed information.

The National Council of Architectural Registration Boards (NCARB) is a nonprofit federation of architectural licensing boards. It has some very useful programs, such as IDP, to assist you in obtaining your architect's license. Contact NCARB for detailed information.

Back Page Promotion

You may be interested in some other books written by Gang Chen:

A. **ARE Mock Exam series.** See the following link:
   http://www.GreenExamEducation.com

B. **LEED Exam Guides series.** See the following link:
   http://www.GreenExamEducation.com

C. ***Building Construction:*** *Project Management, Construction Administration, Drawings, Specs, Detailing Tips, Schedules, Checklists, and Secrets Others Don't Tell You (Architectural Practice Simplified, 2nd edition)*
   http://www.GreenExamEducation.com

D. ***Planting Design Illustrated***
   http://www.GreenExamEducation.com

ARE Mock Exam Series

Published ARE books (One Mock Exam book for each ARE division, plus California Supplemental Mock Exam):

ARE 5.0 Mock Exam Series
Project Planning & Design (PPD) ARE 5.0 Mock Exam (Architect Registration Examination): ARE 5.0 Overview, Exam Prep Tips, Hot Spots, Case Studies, Drag-and-Place, Solutions and Explanations. **ISBN:** 9781612650296

Other books in the ARE 5.0 Mock Exam Series are being produced. Our goal is to produce one mock exam book PLUS one guidebook for each of the ARE 5.0 exam divisions.

ARE 4.0 Mock Exam Series
Programming, Planning & Practice (PPP) ARE Mock Exam (Architect Registration Examination): ARE Overview, Exam Prep Tips, Multiple-Choice Questions and Graphic Vignettes, Solutions and Explanations. **ISBN-13:** 9781612650067

Site Planning & Design ARE Mock Exam (SPD of Architect Registration Examination): ARE Overview, Exam Prep Tips, Multiple-Choice Questions and Graphic Vignettes, Solutions and Explanations. **ISBN-13:** 9781612650111

Building Design and Construction Systems (BDCS) ARE Mock Exam (Architect Registration Examination): ARE Overview, Exam Prep Tips, Multiple-Choice Questions and Graphic Vignettes, Solutions and Explanations. **ISBN-13:** 9781612650029

Schematic Design (SD) ARE Mock Exam (Architect Registration Examination): ARE Overview, Exam Prep Tips, Graphic Vignettes, Solutions and Explanations
ISBN: 9781612650050

Structural Systems ARE Mock Exam (SS of Architect Registration Examination): ARE Overview, Exam Prep Tips, Multiple-Choice Questions and Graphic Vignettes, Solutions and Explanations. **ISBN:** 9781612650012

Building Systems (BS) ARE Mock Exam (Architect Registration Examination): ARE Overview, Exam Prep Tips, Multiple-Choice Questions and Graphic Vignettes, Solutions and Explanations. **ISBN-13:** 9781612650036

Construction Documents and Service (CDS) Are Mock Exam (Architect Registration Examination): ARE Overview, Exam Prep Tips, Multiple-Choice Questions and Graphic Vignettes, Solutions and Explanations. **ISBN-13:** 9781612650005

Mock California Supplemental Exam (CSE of Architect Registration Examination): CSE Overview, Exam Prep Tips, General Section and Project Scenario Section, Questions, Solutions and Explanations. **ISBN:** 9781612650159

Upcoming ARE books:
Other books in the ARE Mock Exam Series are being produced. Our goal is to produce one mock exam book PLUS one guidebook for each of the ARE exam divisions.

See the following link for the latest information:
http://www.GreenExamEducation.com

LEED Exam Guides series*:* Comprehensive Study Materials, Sample Questions, Mock Exam, Building LEED Certification and Going Green

LEED (Leadership in Energy and Environmental Design) is the most important trend of development, and it is revolutionizing the construction industry. It has gained tremendous momentum and has a profound impact on our environment.

From LEED Exam Guides series, you will learn how to

1. Pass the LEED Green Associate Exam and various LEED AP + exams (each book will help you with a specific LEED exam).

2. Register and certify a building for LEED certification.

3. Understand the intent for each LEED prerequisite and credit.

4. Calculate points for a LEED credit.

5. Identify the responsible party for each prerequisite and credit.

6. Earn extra credit (exemplary performance) for LEED.

7. Implement the local codes and building standards for prerequisites and credit.

8. Receive points for categories not yet clearly defined by USGBC.

There is currently NO official book on the LEED Green Associate Exam, and most of the existing books on LEED and LEED AP are too expensive and too complicated to be practical and helpful. The pocket guides in LEED Exam Guides series fill in the blanks, demystify LEED, and uncover the tips, codes, and jargon for LEED as well as the true meaning of "going green." They will set up a solid foundation and fundamental framework of LEED for you. Each book in the LEED Exam Guides series covers every aspect of one or more specific LEED rating system(s) in plain and concise language and makes this information understandable to all people.

These pocket guides are small and easy to carry around. You can read them whenever you have a few extra minutes. They are indispensable books for all people—administrators; developers; contractors; architects; landscape architects; civil, mechanical, electrical, and plumbing engineers; interns; drafters; designers; and other design professionals.

Why is the LEED Exam Guides series needed?

A number of books are available that you can use to prepare for the LEED exams:

1. *USGBC Reference Guides*. You need to select the correct version of the *Reference Guide* for your exam.

   The *USGBC Reference Guides* are comprehensive, but they give too much information. For example, *The LEED 2009 Reference Guide for Green Building Design and Construction (BD&C)* has about 700 oversized pages. Many of the calculations in the books are too detailed for the exam. They are also expensive (approximately $200 each, so most people may not buy them for their personal use, but instead, will seek to share an office copy).

   It is good to read a reference guide from cover to cover if you have the time. The problem is not too many people have time to read the whole reference guide. Even if you do read the whole guide, you may not remember the important issues to pass the LEED exam. You need to reread the material several times before you can remember much of it.

   Reading the reference guide from cover to cover without a guidebook is a difficult and inefficient way of preparing for the LEED AP Exam, because you do NOT know what USGBC and GBCI are looking for in the exam.

2. The USGBC workshops and related handouts are concise, but they do not cover extra credits (exemplary performance). The workshops are expensive, costing approximately $450 each.

3. Various books published by a third party are available on Amazon, bn.com and books.google.com. However, most of them are not very helpful.

   There are many books on LEED, but not all are useful.

   LEED Exam Guides series will fill in the blanks and become a valuable, reliable source:

   a. They will give you more information for your money. Each of the books in the LEED Exam Guides series has more information than the related USGBC workshops.

   b. They are exam-oriented and more effective than the USGBC reference guides.

   c. They are better than most, if not all, of the other third-party books. They give you comprehensive study materials, sample questions and answers, mock exams and answers, and critical information on building LEED certification and going green. Other third-party books only give you a fraction of the information.

   d. They are comprehensive yet concise. They are small and easy to carry around. You can read them whenever you have a few extra minutes.

   e. They are great timesavers. I have highlighted the important information that you need to understand and MEMORIZE. I also make some acronyms and short sentences to help you easily remember the credit names.

It should take you about 1 or 2 weeks of full-time study to pass each of the LEED exams. I have met people who have spent 40 hours to study and passed the exams.

You can find sample texts and other information on the LEED Exam Guides series in customer discussion sections under each of my book's listing on Amazon, bn.com and books.google.com.

What others are saying about *LEED GA Exam Guide* (Book 2, LEED Exam Guide series):

"Finally! A comprehensive study tool for LEED GA Prep!

"I took the 1-day Green LEED GA course and walked away with a power point binder printed in very small print—which was missing MUCH of the required information (although I didn't know it at the time). I studied my little heart out and took the test, only to fail it by 1 point. Turns out I did NOT study all the material I needed to in order to pass the test. I found this book, read it, marked it up, retook the test, and passed it with a 95%. Look, we all know the LEED GA exam is new and the resources for study are VERY limited. This one is the VERY best out there right now. I highly recommend it."
—**ConsultantVA**

"Complete overview for the LEED GA exam

"I studied this book for about 3 days and passed the exam … if you are truly interested in learning about the LEED system and green building design, this is a great place to start."
—**K.A. Evans**

"A Wonderful Guide for the LEED GA Exam

"After deciding to take the LEED Green Associate exam, I started to look for the best possible study materials and resources. From what I thought would be a relatively easy task, it turned into a tedious endeavor. I realized that there are vast amounts of third-party guides and handbooks. Since the official sites offer little to no help, it became clear to me that my best chance to succeed and pass this exam would be to find the most comprehensive study guide that would not only teach me the topics, but would also give me a great background and understanding of what LEED actually is. Once I stumbled upon Mr. Chen's book, all my needs were answered. This is a great study guide that will give the reader the most complete view of the LEED exam and all that it entails.

"The book is written in an easy-to-understand language and brings up great examples, tying the material to the real world. The information is presented in a coherent and logical way, which optimizes the learning process and does not go into details that will not be needed for the LEED Green Associate Exam, as many other guides do. This book stays dead on topic and keeps the reader interested in the material.

"I highly recommend this book to anyone that is considering the LEED Green Associate Exam. I learned a great deal from this guide, and I am feeling very confident about my chances for passing my upcoming exam."
—**Pavel Geystrin**

"Easy to read, easy to understand

"I have read through the book once and found it to be the perfect study guide for me. The author does a great job of helping you get into the right frame of mind for the content of the exam. I had started by studying the Green Building Design and Construction reference guide for LEED projects produced by the USGBC. That was the wrong approach, simply too much information with very little retention. At 636 pages in textbook format, it would have been a daunting task to get through it. Gang Chen breaks down the points, helping to minimize the amount of information but maximizing the content I was able to absorb. I plan on going through the book a few more times, and I now believe I have the right information to pass the LEED Green Associate Exam."
—**Brian Hochstein**

"All in one—LEED GA prep material

"Since the LEED Green Associate exam is a newer addition by USGBC, there is not much information regarding study material for this exam. When I started looking around for material, I got really confused about what material I should buy. This LEED GA guide by Gang Chen is an answer to all my worries! It is a very precise book with lots of information, like how to approach the exam, what to study and what to skip, links to online material, and tips and tricks for passing the exam. It is like the 'one stop shop' for the LEED Green Associate Exam. I think this book can also be a good reference guide for green building professionals. A must-have!"
—**SwatiD**

"An ESSENTIAL LEED GA Exam Reference Guide

"This book is an invaluable tool in preparation for the LEED Green Associate (GA) Exam. As a practicing professional in the consulting realm, I found this book to be all-inclusive of the preparatory material needed for sitting the exam. The information provides clarity to the fundamental and advanced concepts of what LEED aims to achieve. A tremendous benefit is the connectivity of the concepts with real-world applications.

"The author, Gang Chen, provides a vast amount of knowledge in a very clear, concise, and logical media. For those that have not picked up a textbook in a while, it is very manageable to extract the needed information from this book. If you are taking the exam, do yourself a favor and purchase a copy of this great guide. Applicable fields: Civil Engineering, Architectural Design, MEP, and General Land Development."
—**Edwin L. Tamang**

Note: Other books in the **LEED Exam Guides series** are in the process of being produced. At least **one book will eventually be produced for each of the LEED exams.** The series include:

LEED v4 Green Associate Exam Guide (LEED GA): Comprehensive Study Materials, Sample Questions, Mock Exam, Green Building LEED Certification, and Sustainability, LEED Exam Guide series, ArchiteG.com. Latest Edition.

LEED GA MOCK EXAMS (LEED v4): *Questions, Answers, and Explanations: A Must-Have for the LEED Green Associate Exam, Green Building LEED Certification, and Sustainability*, LEED Exam Guide series, ArchiteG.com. Latest Edition

LEED v4 BD&C EXAM GUIDE: *A Must-Have for the LEED AP BD+C Exam: Comprehensive Study Materials, Sample Questions, Mock Exam, Green Building Design and Construction, LEED Certification, and Sustainability*, LEED Exam Guide series, ArchiteG.com. Latest Edition.

LEED v4 BD&C MOCK EXAMS: *Questions, Answers, and Explanations: A Must-Have for the LEED AP BD+C Exam, Green Building LEED Certification, and Sustainability*, LEED Exam Guide series, ArchiteG.com. Latest Edition.

LEED ID&C Exam Guide: *A Must-Have for the LEED AP ID+C Exam: Study Materials, Sample Questions, Green Interior Design and Construction, Green Building LEED Certification, and Sustainability*, LEED Exam Guide series, ArchiteG.com. Latest Edition.

LEED ID&C Mock Exam: *Questions, Answers, and Explanations: A Must-Have for the LEED AP ID+C Exam, Green Interior Design and Construction, Green Building LEED Certification, and Sustainability*, LEED Exam Guide series, ArchiteG.com. Latest Edition.

LEED O&M MOCK EXAMS: *Questions, Answers, and Explanations: A Must-Have for the LEED O&M Exam, Green Building LEED Certification, and Sustainability*, LEED Exam Guide series, ArchiteG.com. Latest Edition.

LEED O&M EXAM GUIDE: *A Must-Have for the LEED AP O+M Exam: Comprehensive Study Materials, Sample Questions, Mock Exam, Green Building Operations and Maintenance, LEED Certification, and Sustainability*, LEED Exam Guide series, ArchiteG.com. Latest Edition.

LEED HOMES EXAM GUIDE: *A Must-Have for the LEED AP Homes Exam: Comprehensive Study Materials, Sample Questions, Mock Exam, Green Building LEED Certification, and Sustainability*, LEED Exam Guide series, ArchiteG.com. Latest Edition.

LEED ND EXAM GUIDE: *A Must-Have for the LEED AP Neighborhood Development Exam: Comprehensive Study Materials, Sample Questions, Mock Exam, Green Building LEED Certification, and Sustainability*, LEED Exam Guide series, ArchiteG.com. Latest Edition.

How to order these books:
You can order the books listed above at:
http://www.GreenExamEducation.com

OR
http://www.ArchiteG.com

Building Construction

Project Management, Construction Administration, Drawings, Specs, Detailing Tips, Schedules, Checklists, and Secrets Others Don't Tell You (Architectural Practice Simplified, 2nd edition)

Learn the Tips, Become One of Those Who Know Building Construction and Architectural Practice, and Thrive!

For architectural practice and building design and construction industry, there are two kinds of people: those who know, and those who don't. The tips of building design and construction and project management have been undercover—until now.

Most of the existing books on building construction and architectural practice are too expensive, too complicated, and too long to be practical and helpful. This book simplifies the process to make it easier to understand and uncovers the tips of building design and construction and project management. It sets up a solid foundation and fundamental framework for this field. It covers every aspect of building construction and architectural practice in plain and concise language and introduces it to all people. Through practical case studies, it demonstrates the efficient and proper ways to handle various issues and problems in architectural practice and building design and construction industry.

It is for ordinary people and aspiring young architects as well as seasoned professionals in the construction industry. For ordinary people, it uncovers the tips of building construction; for aspiring architects, it works as a construction industry survival guide and a guidebook to shorten the process in mastering architectural practice and climbing up the professional ladder; for seasoned architects, it has many checklists to refresh their memory. It is an indispensable reference book for ordinary people, architectural students, interns, drafters, designers, seasoned architects, engineers, construction administrators, superintendents, construction managers, contractors, and developers.

You will learn:
1. How to develop your business and work with your client.
2. The entire process of building design and construction, including programming, entitlement, schematic design, design development, construction documents, bidding, and construction administration.
3. How to coordinate with governing agencies, including a county's health department and a city's planning, building, fire, public works departments, etc.
4. How to coordinate with your consultants, including soils, civil, structural, electrical, mechanical, plumbing engineers, landscape architects, etc.
5. How to create and use your own checklists to do quality control of your construction documents.
6. How to use various logs (i.e., RFI log, submittal log, field visit log, etc.) and lists (contact list, document control list, distribution list, etc.) to organize and simplify your work.
7. How to respond to RFI, issue CCDs, review change orders, submittals, etc.
8. How to make your architectural practice a profitable and successful business.

Planting Design Illustrated
A Must-Have for Landscape Architecture: A Holistic Garden Design Guide with Architectural and Horticultural Insight, and Ideas from Famous Gardens in Major Civilizations

One of the most significant books on landscaping!

This is one of the most comprehensive books on planting design. It fills in the blanks of the field and introduces poetry, painting, and symbolism into planting design. It covers in detail the two major systems of planting design: formal planting design and naturalistic planting design. It has numerous line drawings and photos to illustrate the planting design concepts and principles. Through in-depth discussions of historical precedents and practical case studies, it uncovers the fundamental design principles and concepts, as well as the underpinning philosophy for planting design. It is an indispensable reference book for landscape architecture students, designers, architects, urban planners, and ordinary garden lovers.

What Others Are Saying about *Planting Design Illustrated* ...

"I found this book to be absolutely fascinating. You will need to concentrate while reading it, but the effort will be well worth your time."
—**Bobbie Schwartz, former president of APLD (Association of Professional Landscape Designers) and author of** *The Design Puzzle: Putting the Pieces Together*.

"This is a book that you have to read, and it is more than well worth your time. Gang Chen takes you well beyond what you will learn in other books about basic principles like color, texture, and mass."
—**Jane Berger, editor & publisher of gardendesignonline**

"As a longtime consumer of gardening books, I am impressed with Gang Chen's inclusion of new information on planting design theory for Chinese and Japanese gardens. Many gardening books discuss the beauty of Japanese gardens, and a few discuss the unique charms of Chinese gardens, but this one explains how Japanese and Chinese history, as well as geography and artistic traditions, bear on the development of each country's style. The material on traditional Western garden planting is thorough and inspiring, too. *Planting Design Illustrated* definitely rewards repeated reading and study. Any garden designer will read it with profit."
—**Jan Whitner, editor of the** *Washington Park Arboretum Bulletin*

"Enhanced with an annotated bibliography and informative appendices, *Planting Design Illustrated* offers an especially "reader friendly" and practical guide that makes it a very strongly recommended addition to personal, professional, academic, and community library gardening & landscaping reference collection and supplemental reading list."
—**Midwest Book Review**

"Where to start? *Planting Design Illustrated* is, above all, fascinating and refreshing! Not something the lay reader encounters every day, the book presents an unlikely topic in an easily digestible, easy-to-follow way. It is superbly organized with a comprehensive table of contents, bibliography, and appendices. The writing, though expertly informative, maintains its accessibility throughout and is a joy to read. The detailed and beautiful illustrations expanding on the concepts presented were my favorite portion. One of the finest books I've encountered in this contest in the past 5 years."
—Writer's Digest 16th Annual International Self-Published Book Awards Judge's Commentary

"The work in my view has incredible application to planting design generally and a system approach to what is a very difficult subject to teach, at least in my experience. Also featured is a very beautiful philosophy of garden design principles bordering poetry. It's my strong conviction that this work needs to see the light of day by being published for the use of professionals, students & garden enthusiasts."
—Donald C. Brinkerhoff, FASLA, chairman and CEO of Lifescapes International, Inc.

Index

3016 rule, 14, 31, 33
A/E, 21, 37, 38
accessibility issues, 79, 124, 129
acoustical lining, 63, 74, 106, 118
ADAAG, 77
AGS, 106, 107
AIA, 16, 166, 168
air chamber, 60, 105
albedo, 73, 116
Aluminum, 78, 123
ARE Mock Exam, 3, 7, 9, 13, 15, 16, 30, 45, 97, 169, 170, 171
ARE Guidelines, 19
ASCE, 113, 114
AXP, 13, 17, 19, 20, 22, 23
AXP Portfolio, 20
baffles, 74, 118
Baked enamel, 80, 126
Biophilia, 87, 132
Bituminous Systems, 124, 125
Bogs, 73, 117
breaks, 3, 14, 34, 176
Brick pattern, 56
Built Up Roof, 79, 124
capillary action, 127
case studies, 9, 11, 17, 178, 179
CE, 14, 17, 25, 152, 153, 154, 155
centroid, 110
centroidal axis, 110
check, 20
check-all-that-apply, 11, 17
CMU, 58, 78, 102, 121, 122
Codes and standards, 14, 35
<u>**coefficient of utilization**</u>, 74
condensation, 103, 119
Conditional Use Permit, 73, 116
conduction, 106
conductivity, 73, 116

construction joint, 99
convection, 106
convex slope, 49, 98
CSI, 42
Dead loads, 97
Detention ponds, 86, 132
dew point, 103
direction of water flow, 46, 97
distracters, 117
double duct system, 59, 104
drag-and-place, 11, 17
drip edge, 99
Dry ice, 84, 130
Duct silencers, 74, 118
Dynamic loads, 97
easement, 74, 117
English system, 34
equinox, 73, 116
erosion control, 86
Erosion control, 131
exam content, 23, 29, 37
Exam Format & Time, 24
FAR, 74, 117
FCU, 76, 120
FEMA454, 15, 41
Fens, 73, 117
finish schedules, 78, 122
Floor Area Allowances, 57
free-body diagram, 111, 112
FRP, 78, 122
GeeForum.com, 138
graywater, 72, 115
headroom, 78, 121, 122
Health Department, 78, 122
historic building, 83, 129, 130
HOA, 87, 132
horizontal exit, 79, 123
hot spots, 11, 17

Hot-dipped galvanized steel, 78
IDP, 13, 17, 19, 21, 22, 26, 162, 168
Impact loads, 97
Incentive Zoning, 73, 116
inch-pound units, 34
Index, 181
Insulated Concrete Forms (ICF) system, 53, 99
intermediate landing, 58, 102
intern, 22
International Building Code (IBC), 39, 45, 97
interns, 22
invert elevation, 50, 98
LCA, 87, 133
lead-based paint, 83, 129, 130
LEED, 16, 169, 173, 174, 175, 176, 177
LID, 87, 133
Life cycle analysis, 133
Life cycle approach, 133
Life cycle cost and saving analysis, 133
life cycle costing, 87, 133
lumens, 74, 75, 118
lux, 118
Marshes, 73, 117
masonry veneers, 56, 100
MEEB, 107
membrane roofing systems, 124
metric system, 34
mnemonic, 31
mnemonics, 14, 31, 32
Mnemonics, 31, 32, 42
moment of inertia, 70, 110
Mortise lock, 51, 98
mullion, 53, 99
multiple choice, 11, 17
NCARB, 29
NCARB **Rules of Conduct**, 125
Non-conforming Use, 73, 116
Northern Hemisphere, 67, 73, 116
Note, 27, 38, 40, 41, 43, 84, 85, 97, 99, 100, 101, 104, 108, 112, 114, 122, 127, 165, 176
Occupancy group, 114, 115
occupant load, 126
Orientation, 86, 131

PA, 14, 17, 152, 153, 154, 155
panel schedules, 75, 119
panic hardware, 56, 80, 101, 125, 126
passing or failing percentage, 25
PcM, 14, 17, 152, 153, 154, 155
PDD, 14, 17, 28, 152, 153, 154, 155
physical exercise, 14, 34
pivoting window, 98
PjM, 14, 17, 152, 153, 154, 155
power plans, 75, 119
PPD, 1, 2, 7, 11, 14, 15, 17, 37, 38, 45, 97, 138, 152, 153, 154, 155, 170
preformed synthetic rubber, 84, 130
PRV, 76, 120
PVC, 79, 124, 125
quantitative fill-in-the-blank, 11, 17
radiation, 106
Radon gas, 80, 125
reactions, 67, 108, 109, 111
Refrigeration Flow Diagram, 61, 106
register, 24, 25
relative humidity, 58, 103
reporting hours, 20
Restrictive covenants, 87, 132
Retention ponds, 86, 132
rolling clock, 23, 24
roof overhang, 86, 131
routine, 14, 33
Rules of Conduct, 22
R-value, 65, 77, 107
scores, 25
section, 42
sediment control, 86, 131
Seismic Design Category, 72, 114, 115
separation distance, 79, 123
sewer line, 50, 98
shock absorber, 60, 105
shop drawings, 121
SI units, 34
Single Ply Roof Systems, 124
single zone system, 59, 104
single-line diagram, 75, 119
site coverage ratio, 74, 117
Six-Month Rule, 20
smoke, 77, 121
solstice, 73, 116

spot elevation, 98
stack bond, 56, 100
Stress, 40, 156
surface roughness, 72, 112
Swamps, 73, 117
table 3.1, 71, 111
tension, 111
terminal reheat system, 105
test-taking tips, 25, 26
Tips, 169, 178
ton, 58, 104

U/A, 21, 37
U-value, 65, 77, 107, 120
variable flow system, 59, 104
vibration, 83, 130
vignette section, 19
W shape steel, 57, 102
water resistant membrane, 85, 130
Waterstop, 84, 130
Weather Struck brick joint, 82, 128
Wiring Diagram, 62, 64
zoning, 74, 116, 117

Made in the USA
Monee, IL
02 March 2020